RA
395
S3
S5
1970

WITHDRAWN 71-566

Shine
Serendipity in St.
Helena.

JUL 2000

Date Due

		JUN	2004
		JUN 09	
		JUL X X 2015	

SERENDIPITY IN ST. HELENA

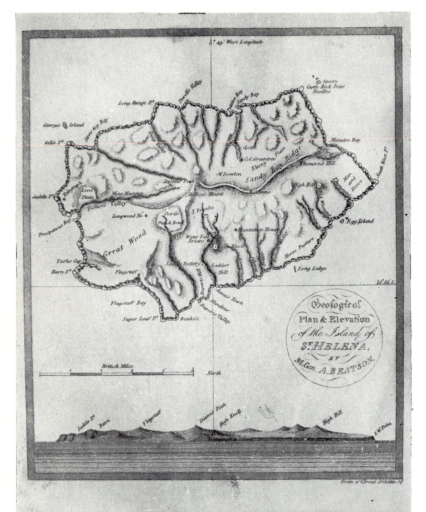

Geological Plan & Elevation of the Island of St HELENA, BY M.Gen. A.BEATSON.

SERENDIPITY IN ST. HELENA

A GENETICAL AND MEDICAL STUDY
OF AN ISOLATED COMMUNITY

IAN SHINE, M.D.

with the assistance of
REYNOLD GOLD, M.B.

and a foreword by
PROFESSOR SIR MAX ROSENHEIM, P.R.C.P.

PERGAMON PRESS
Oxford · New York
Toronto · Sydney · Braunschweig

Pergamon Press Ltd., Headington Hill Hall, Oxford

Pergamon Press Inc., Maxwell House, Fairview Park, Elmsford,
New York 10523

Pergamon of Canada Ltd., 207 Queen's Quay West, Toronto 1

Pergamon Press (Aust.) Pty. Ltd., 19a Boundary Street,
Rushcutters Bay, N.S.W. 2011, Australia

Vieweg & Sohn GmbH, Burgplatz 1, Braunschweig

First edition 1970

Library of Congress Catalog Card No. 68–8533

Printed in Great Britain by A. Wheaton & Co., Exeter

08 012794 0

CONTENTS

LIST OF ILLUSTRATIONS

LIST OF TABLES

FOREWORD

WHILE doing his National Service in the Army, Dr. Ian Shine volunteered to go out to the island of St. Helena as medical officer. He became intrigued and interested by the variety of genetic anomalies that he found in his patients and, in the course of two tours of duty, he carried out a genetic survey of the population. This is what this book records.

Serendipity has been defined as the faculty and habit of making felicitous discoveries by accident and the term derives from a fairy-tale about the island of Ceylon. It was certainly a happy and fortunate chance that landed Dr. Shine on St. Helena, but his survey of most of the 5000 inhabitants and the careful recording of his findings were certainly not matters of chance. I have followed Dr. Shine's work with great interest, have read his book with real enjoyment and regard it as a privilege to have been invited to contribute a foreword. It has been fascinating to see how his decision to volunteer has determined his future career. Originally destined for general practice, his clinical observations, his interest and drive, and his subsequent training at the Medical Research Council's Population Genetics Research Unit in Oxford, have made him a true clinical geneticist.

When the inhabitants of Tristan da Cunha were evacuated to England following the volcanic eruption, they were extensively examined and investigated by a team of experts. I believe that the findings recorded here by a single doctor who managed to examine almost all the inhabitants of St. Helena on the spot, provide material of at least equal interest. There is something of the Swiss Family Robinson in Dr. Shine's saga, for he had to use considerable ingenuity not only in obtaining information from his patients, but also in devising methods of measurement and investigation. This personal account of his survey makes a most fascinating story.

M. L. ROSENHEIM

ACKNOWLEDGEMENTS

IT IS inevitable that a study taking 6 years to complete should leave me much indebted to a great many colleagues, friends and 4500 St. Helenians whose diseases and peculiarities receive more emphasis than their sense of humour, kindness and generosity. Their present disposition fits well with the description given of them two centuries ago by Captain James Cook.

After 6 months' residence on the island I reported my findings to Dr. H. N. Levitt, asking for his opinion, and was gratified to learn that he found them interesting. This was valuable advice without which I would probably have abandoned the study and ascribed the unique findings to a fertile imagination flowering under complete isolation.

In the conduct of the investigation I received valuable assistance from successive Governors, Sir Robert Alford and Sir John Field, and from their governments. I am also grateful to the heads of departments, Messrs. Thorpe & Son; Solomon & Company, Cable and Wireless Ltd.; and the U.S. Military Commander on Ascension Island for allowing me to examine their employees. I would like to emphasize how much the island appreciated the generosity of the United States. When a geiger counter was required they supplied one within 2 days, when we ran out of antibiotics, they sent a ship about 1400 miles with a free gift of six months' supply, and when, in 1963, there was an insufficient medical staff to cope with a series of emergencies, they offered to parachute a medical team onto the island.

In the collection of data I received invaluable assistance from the following, to whom I am most grateful: Miss Dianna Dillon, Mrs. Joan Johns, Mr. D. Cribbs, Dr. T. Geddes, Mr. A. Richards, Mrs. E. Frater, Mr. A. Loveridge, Mrs. Bronwyn McLeary and Mr. E. Constantine. I am also very grateful to the hospital staff, headed by Miss Grace Sim, for their willing co-operation. I would like to thank Miss Edith Johns who was a font of knowledge and a valuable friend. After serving for 9 months single-handed, I was relieved by Dr. J. S. Noaks who generously undertook additional routine duties and so enabled me to devote further time to this study.

I would like to thank the Eugenics Society who provided £200 for the purchase of books and equipment, the Medical Research Council Population Genetics Research Unit who loaned a transistorized electrocardiogram which was invaluable as the country districts were without electricity, and Professor R. Turner of the Government Pathological Laboratory at Cape Town, who generously donated several litters of mice, analysed samples of water and aqueduct for arsenic, performed histological examination of postmortem material and generously shouldered many of St. Helena's problems.

In medical matters I was privileged to have the opportunity to discuss several problems with eminent experts. Advice was given by Dr. J. M. Cullen, Dr. N. J. Bailey and Mr. J. S. Scott; Dr. A. Barr and Miss R. Ash on computing, programming and statistics; Dr. M. Balint on the "evil eye"; Professor G. Belyavin and Dr. A. Cohen on virology; Dr. Barbara Billing on hyperbilirubinaemia; Dr. Rosemary Biggs and Professor R. G. Macfarlane on blood coagulation who also undertook factor IX assay. Also Professor C. E. Dent, Professor G. Rimmington, Dr. E. B. Dowdle and Dr. E. J. Ross were consulted on metabolic diseases; Dr. J. Gear kindly undertook viral studies; Dr. P. J. Hare, Dr. C. F. H. Vickers and Dr. R. S. Wells gave advice on dermatology; Dr. Catherine Hollman on hallux valgus; Professor J. N. Morris and Dr. W. E. Miall on ischaemic heart disease; Professor Sir George Pickering on arterial disease; Sir Herbert Seddon on orthopaedics; Professor J. M. Tanner on dwarfism; Dr. W. R. Trotter on the thyroid; Professor J. Yudkin on nutrition; and Professor A. W. Woodruff on tropical medicine. I am deeply grateful for their help.

I also wish to thank the Royal Air Force for flying blood samples from Ascension Island to England.

I am greatly indebted to Dr. A. C. Stevenson for giving me instruction in genetics, generous assistance and much useful criticism of the manuscript.

I would like to express my appreciation for the stimulus of Professor N. E. Morton who has made many useful suggestions about the text.

The first chapter is based on a book that Dr. R. Gold and I

began writing about the St. Helenians, but stopped on reflecting that I had collected the data as their private doctor. I am much indebted to Dr. Gold for applying his remarkable skills to my disjointed account and then generously allowing me to use the material here.

These pages have been expertly typed more than once by Mrs. Mary Parke, Mrs. Yoshi Kaneshiro and Mrs. Freda Hellinger; in addition, Mrs. Parke kindly performed many of the calculations and tabulations.

As this work forms a reasonably complete account of a single study, it was decided to publish it in one place rather than break it up into several papers. Some of these data appeared in an M.D. thesis at Cambridge whose Regius Professor has kindly allowed me to use them here.

Finally, I would like to offer my warm thanks to Sir Max Rosenheim for his expert assistance and friendly advice.

I.S.

As to genius and temper of these people, they seemed to us the most honest, the most inoffensive and the most hospitable people we had ever met with of English extraction, having scarse any tincture of avarice or ambition. He asks some if they have no curiosity to see the rest of the world, and how they could confine themselves to so small a spot of earth separated at such distance from mankind. They replied that they enjoyed the necessaries of life in great plenty, they were neither parched with excessive heat or pinched with cold: they lived in perfect security, in no danger of enemies, of robbers, of wild beasts or rigorous seasons, and were happy in the enjoyment of a continued state of health.

(CAPTAIN COOK, F.R.S., 1775)

CHAPTER I

INTRODUCTION

Let us pause to consider the English,
Who when they pause to consider themselves they get all reticently thrilled and
   tinglish.
Englishmen are distinguished by their traditions and ceremonials,
And also by their affection for their colonies and their condescension to their
   colonials.

(Ogden Nash)

In January 1960 a notice appeared in daily orders in all units of the Armed Forces asking for volunteers to go to the island of St. Helena where the only doctor was due to leave by the next boat. Believing St. Helena to be in the Pacific Ocean which I had always hoped to visit, I asked Colonel Ian Milne, my commanding officer, for permission to volunteer; he agreed and 3 weeks later I sailed from Southampton. During the subsequent $2\frac{1}{2}$ years as the island medical officer, I came across several unique syndromes and some fascinating research problems, but as I had neither the training nor the facilities for research, the problems were studied in a makeshift manner, relying on simple clinical observations with few laboratory investigations. The purpose of this report is to give an account of these problems and their solution, together with a description of the circumstances under which the work was done, for without the flavour of the field situation it is hard to gauge the validity of the observations.

According to Sir Karl Popper (1963), "The belief that we can start with observations alone, without anything in the nature of a theory is absurd". He believed that scientific research starts with a hypothesis, which is then tested by experiment the results of which prove or disprove the original predictions. This orderly sequence was not followed in this study which started with the chance observations of rare genetical abnormalities. Because of these observations the islanders were examined to measure the prevalence of local disease; and dietary, demographic, medical and pedigree

1

data were collected in an attempt to understand the nature of the natural circumstances that might be responsible. The hypothesis that attributes these genetical abnormalities to the island's physical isolation emerged long after all the data were collected while attempting to evaluate environmental and hereditary influences on the local disease pattern.

THE NATURE–NURTURE PROBLEM

Even though the conflict between nature and nurture has lost the force that it held for Pearson and the early geneticists, it yet remains a central theme in biology, and one that was evaluated in relation to the data collected in St. Helena. The sharp distinction depends largely on semantics, particularly on the time-scale over which the causal agent is considered to act, or, as Haldane (1936) expressed it, "the distinction between organism and environment is an abstraction that the geneticist finds hard to make in time". This is now generally conceded, although it is easy to adopt a limited viewpoint—forgetting that phenylketonuria is a genetical disease provided everyone eats phenylalanine, but environmental if only some people do; likewise, as Snyder (1959) pointed out, scurvy is an environmental disease provided everyone lacks L-gulonolactone dehydrogenase, but genetical if only some do. Even eminent men have occasionally regarded the demonstration that a disease was environmentally determined as evidence that it was not inherited and vice versa, suggesting that the old dichotomy still has its admirers. In his campaign against the abolitionists, Sir Ronald Fisher (1959a) cited the twin evidences of genetical predisposition to smoking as an indication that the products of tobacco combustion were not carcinogenic. Nowadays, it is to be hoped that no one would regard the discovery of genetically determined variation in sensitivity to tobacco carcinogens as reason for doubting their lethality, any more than one would doubt that the Bittner milk factor is carcinogenic to mice merely because some strains of mice are more sensitive than others.

There are many well-established experiments that demonstrate the dependence of gene expression upon environmental conditions as well as examples where gene expression is modified by the presence of other genes. The mutant gene that fails to code tyrosinase will cause albinism only if the allelic gene is also mutant; and the

2

ABO phenotype will only be expressed in the presence of the wild-type gene at the X locus (Bombay supressor—Levine *et al.*, 1955). The Himalayan rabbit experiment provides the most picturesque demonstration of environmental modification of gene effect. This rabbit inherits fur that is white except on the paws, head and tail, where it is black. If the skin temperature is lowered, as it is at the extremities, or if the trunk is shaved, an allele of the albino locus produces tyrosinase, and black fur is produced in this cold patch, but if the patch is kept warm during regeneration, the new hair is white instead of black (Engelsmeier, 1937; Danneel, 1938).

The evolution of ideas about the aetiology of leprosy and mongolism illustrates the conceptual difficulty to which Haldane referred. Leprosy has been widely considered infectious for many centuries. By contrast, Danielssen and Boeck (1848) suggested that the disease was inherited, and this view was supported by Sedgwick. However, when Hansen showed that leprosy was due to *Mycobacterium leprae* in 1874, the hypothesis that leprosy was inherited fell into disrepute. Even Penrose (1934) stated that "as the causal bacillus has been discovered, Sedgwick's views made peculiar reading". However, there is now evidence of a genetically determined variation in sensitivity to the mycobacterium (Aycock & Gordon, 1941; Spickett, 1962; Blumberg, 1965). Similar reasoning may be applied to the cause of mongolism which becomes genetically or environmentally determined according to the time scale that is adopted. The abnormality appears to be genetical on the basis of twin concordance and the presence of autosomal trisomy, but only if the present generation of mongols is considered. If causes acting on the parental generation are considered, the maternal age effect suggests the possibility of an environmental determinant, which may be a virus (Stoller & Collmann, 1965). If more distant generations are considered, evidence of familial predisposition to the environmental agent may become apparent, and Penrose (1963) considers that there may be a familial predisposition in 25% of mongols. If it should be established that there is a variation in susceptibility to an environmental determinant, the genes, if any, underlying this susceptibility, like all others depended for their survival upon the bygone environment of early man and ultimately upon mutation: thus, the distinction between nature and nurture becomes artificial.

PRACTICAL CONSIDERATIONS

To unravel the contribution of nature and nurture to human disease, some way of fixing the genotype or the environment is required. The first can be accomplished by comparing the differences between identical and fraternal twins as originally suggested by Galton (1874) on the assumption that genotypic differences do not exist between identical twins, and hence phenotypic differences may reasonably be attributed to environment. This method has been used to demonstrate the innateness of criminal behaviour (Lange, 1931), physique (Newman & Freeman, 1937), neurotic illness (Slater, 1953) and physical disease (Gedda, 1951; Hague & Harvald, 1961). The chief difficulties are that twins share the same intrauterine environment as well as genotype, and the assumption of genetic identity is not always justified as shown by the reports of monozygotic twins with differing sex chromosomes (Turpin & Lejeune, 1965; Shine & Corney, 1966). It also assumes that there are only two types of twin which is an unjustified simplification (Allen, 1965; Edwards, 1966). Perhaps the most severe limitation of this method is due to the scarcity of suitable twin pairs; for instance, there were only two such pairs available in St. Helena.

An alternative approach to the problem is to produce or assume a constant environment and equate phenotypic differences with genotypic differences. This seems a reasonable assumption for major gene traits with full penetrance, if reliable family histories can be obtained; however, if there is no typical segregation pattern, or association with consanguinity, or characteristic clinical or biochemical features, it is impossible to know how much of the observed variation is due to genotypic differences, how much to the chance concatenation of random events and how much to environmental factors that were not standardized.

CHARACTERISTICS PECULIAR TO ISOLATES

Isolates present the geneticist with two favourable aspects. Firstly, they act as natural gene-concentrating devices, and secondly, they produce the kind of social situation that a sympathetic investigator can exploit (or if unsympathetic, abuse). It is not so much that the procedures and questions asked are any different from those asked in a large population, but rather that it is easier to

4

obtain reliable answers. The problems of ascertainment, of obtaining physician's and patient's consent and the compilation of pedigrees are much simpler in a small community, such as St. Helena. There the doctor is well respected (particularly if he is the only doctor), and by sharing the problems of the Governor as well as his subjects, he soon acquires a deep understanding of the past history and the present news. It is therefore easy to engage people in conversation and ask delicate questions without offending, whereas in a large population it would require many years of residence to achieve the same rapport. Not only are people more willing to reveal their ancestry but they are better able to do so, since the compilation of pedigrees is a popular pastime, taking the place of monopoly or chess, and is sometimes just as involved. In addition, it is an advantage to have relatives at hand instead of inaccessible in the next city or county.

The syndrome of familial unilateral knock-knees (described in Chapter 4) illustrates some of these advantages. A few people were seen with this deformity, which they attributed to an injury sustained during adolescence. This seemed a reasonable explanation for a unilateral lesion; and, as it was not known to be inherited, it seemed unexceptional when people claimed to be the only affected member of their family. In a city, a single investigator would be unlikely to encounter several examples of a rare symptomless deformity: and if he did, he would not readily appreciate that they were related.

Finally, isolates offer favourable situations in which to evaluate or even measure the effects of certain environmental factors. Potential pathogens, such as motor-cars, shoes, cigarettes and thalidomide are often introduced slowly and separately, with the amount of each item imported being carefully recorded by the customs and excise department. In a larger community there are so many simultaneous changes that it is much harder to measure the effect of a single one and much easier to find spurious correlations, like the highly significant correlation between the annual divorce rate in Britain and the importation of apples, which Sir Ronald Fisher (1959b) attributed to Professor Udny Yule. St. Helena's first motor-car, a model A Ford, registration number 1, imported in 1929, is still running. As my car was given the number plate 242, less research is required to discover the number of registered cars than would be required in the United States. As the first acquisition of a motor-car or a pair of shoes was a noted event, it was well remembered, hence, retrospective information about these things is probably more reliable than in busy Europe.

METHOD OF STUDY

A. *Questions*

The problems that it seemed feasible to investigate were these:

1. Is there a high incidence of congenital and inherited disease?
2. To what extent can the observed pattern of disease be attributed to major genes revealed by inbreeding?
3. Is there less haemolytic disease of the newborn as would be anticipated under inbreeding? (The hospital matron who had many years of experience believed this to be true.)
4. Are the islanders immune from ischaemic heart disease, and, if so, can it be attributed to the high physical activity?
5. As human feet are adapted to walking without shoes, has the recent introduction of shoes caused foot abnormality?
6. What were the underlying metabolic defects in the unique island syndromes?
7. Why was bronchial asthma so common [as it seems to be in Tristan da Cunha (Black *et al.*, 1963)] and why have no deaths ever been recorded from status asthmaticus?
8. Have the people who had mumps orchitis shown diminished fertility? (In 1952 there was an epidemic which attacked about 4000 individuals with a high complication rate.)
9. To what extent does a small polytypic community mate assortively?
10. Is high blood pressure uncommon as it is on some islands? (Lowenstein, 1961; Maddocks & Lovell, 1962; Cruz-Cooke *et al.*, 1964.)
11. What is the extent of the racial admixture?
12. Why should pterygium be common in a dust-free atmosphere?

As the community was virgin from the point of view of medical research, another investigator would doubtless ask another set of twelve questions, and many that were asked were unresolved. It was clearly desirable to limit the study in time and to concentrate on those problems that could be investigated with limited resources. In practice this meant omitting questions 3, 7, 8, 11 and 12. Questions 9 and 10 will not be further considered here because the data have not yet been analysed.

B. *Definition of population*

In an island as small as St. Helena with only 4500 people, personal examination of everyone was a feasible undertaking, whereas it would be virtually impossible in a large community; and it has the advantage that it overcomes the geneticist's usual reliance upon hospital and clinic records, which are inevitably inaccurate to some extent, since they are the result of the observations of several doctors who seldom share the same diagnostic criteria. The resulting bias is avoided if all the observations are made by a single observer.

The island census of November 1956 was used as a definition of the population. An additional reason for examining everyone rather than a selected sample was that random sampling is liable to introduce errors in the ascertainment of rare traits and, moreover, the selection of some islanders, and not others, for examination might have assumed a sinister significance to them and they might not have co-operated. Moreover, there is no problem of ascertainment bias since this method is equivalent to complete selection, i.e. random sampling of families through the parents, without consideration of the phenotypes of the children (Morton, 1959). Although where sibships are not identified through the parents, the method is equivalent to Morton's truncate selection.

The reports of previous medical officers indicated that between 2000 and 3000 new patients were seen annually in the out-patient clinics and so it seemed reasonable to use the clinics as the major source of material. School children were examined during the annual medical inspection and as 99% of deliveries took place in the hospital, babies were examined within 2 days of delivery. Anyone who was not seen in this way was visited at home or at work.

Approximately 90% of the population was examined. The main reason for failing to examine everyone was lack of the examiner's time, but for this there would have been about 99% success. An attempt was made to see the remaining 10% of the population by postponing the termination of the study for 1 month; but as it frequently took 2 hours of strenuous walking to reach some of the outlying homes, the yield was slow, although the people were uniformly co-operative. The minimum estimates of the prevalence of traits in this sample must be reliable since, even if by chance the trait in question is absent in the unexamined group, the

7

prevalence can only change from X/4259 to X/4642, which is not appreciable. It is unlikely that the sample was biased due to deliberate abstention because people would hardly abstain from an ascertainment of several conditions for the same reason; and in practice there were only four abstentions known. Absence of bias is confirmed by the similarity of age and sex structure between the census and sample populations. (Appendix 1.)

METHOD OF INVESTIGATION

At examination the items set out below were obtained from each person. It was hard to know whether the occasional oblique replies were examples of local humour or deliberate elusiveness.

1. *Surname and Christian names.* Usually only one Christian name was given and there was some reluctance to give more, which is illustrated by the following conversation:

I asked a man "Why is your name John D. Brook?"

"My name is John".
"What does the D stand for?"
"It's all in with it".

2. *Age.* Most people knew and gave their exact age. When there was uncertainty relatives were asked for their estimate. One defect in this study is that the date of birth was not obtained which would have been a very important means of identification, since several islanders often share the same surname and Christian name.

3. *District of residence.*

4. *Marital state.*

5. *Total number of live deliveries.* When replying, some people did not include illegitimate children or those who subsequently died, or those who were married. This was not an attempt at concealment but a reflection of a different method of counting children and a literal interpretation of the question "How many children do you have?". The clue was provided by one woman who replied, "I have seven, and another's off hands but still I own it". The attitude to illegitimacy was generally honest and the biological father was invariably acknowledged if the questions were put in an acceptable way; that is, by clearly accepting local standards as normal. Illegitimacy has risen from 20% of births in 1903 to 30% in 1962.

6. *Number of stillbirths and number of children who died.* People sometimes showed difficulty in recalling stillbirths.

7. *Occupation.* Only current occupation was recorded and, for the purposes of regression analysis, was graded according to the amount of physical effort required.

8. *Social class.* People were arbitrarily divided into four classes. (1) Labourers; (2) craftsmen and skilled workers; (3) clerks and professionals; and (4) self-employed. Wives were given the same index as their husbands.

9. *Exercise habit.* People were asked for a current estimate of the number of miles walked daily.

10. *Duration of wearing shoes.*

11. *Smoking habit.* There was usually no resistance to this question but one man when asked, "How many cigarettes do you smoke a day?" replied, "I mostly smoke at night".

12. The body weight was routinely recorded on a Salter bathroom scale, with the patient dressed.

13. *Blood pressure.* Casual blood pressure readings were taken on the right upper arm using a mercury sphygmomanometer with a 4 by 8 in. cuff. The systolic pressure was taken at phase I Korotkoff, and the diastolic at phase IV.

14. The circumference of the extended upper arm was measured at its mid-point with a silk noil millimetre tape measure.

15. The presence of arcus senilis was recorded. Both complete and incomplete arcus were grouped together for present purposes.

The methods adopted for gathering family data and for the detection of hallux valgus, ischaemic heart disease and congenital and hereditary defects will be described in the relevant sections.

MEDICINE AND RESEARCH IN AN UNDEVELOPED COMMUNITY

Research in St. Helena confronts the investigator with two kinds of problem with which he is not accustomed to deal in his home environment. He faces limited technical resources and, in addition, he has to learn to understand a culture that is very different from his own. In order to interpret correctly the islanders' replies to questions and to gain their co-operation and sympathy with my investigations, it was essential for me to become familiar with their customs, attitudes, beliefs, language and background.

HISTORY

St. Helena was discovered by the Portuguese navigator Juan de Nova Castella on 21 May 1502, the feast day of St. Helena, mother of the Emperor Constantine. He landed at the valley where Jamestown now stands and built a chapel there which gave the valley the name Chapel Valley by which it is still locally known.

The Portuguese kept the discovery of the island a secret, landed goats, hogs, pheasants and partridges, planted fruit trees and vegetables and used it as a port of call for their East Indian fleet. It was discovered by the English in 1588, when Captain Thomas Cavendish called on his voyage around the world. He stayed 12 days and recorded the presence of pheasants, partridges, numerous goats and quantities of fruit in Chapel Valley.

Thereafter the English and Dutch merchant ships called frequently to take on water and supplies and to refit. In 1633 the island was annexed by the Dutch but not occupied; it was seized by the English East India Company in 1659; the Dutch re-took it in 1672 and a year later the English recaptured the island and have held it ever since. Apart from the period of Napoleon's exile, the island had been administered by the East India Company under Royal Charter from 1661 until 1834: the Company was then nationalized and the island became a royal possession.

Local history has been admirably recorded by Brooke (1824), Janisch (1885) and Gosse (1938), and although fascinating, it has little relevance to the outside world which remembers St. Helena only as a prison or as a calling place for some eminent person. In 1676, Halley, then a student at Oxford, arrived to determine the position of the fixed stars in the southern hemisphere. Robert Jenkins, of Jenkins's Ear fame, was Governor from 1741 to 1742. Captain Cook visited the island in 1775, and Charles Darwin stopped off for a few days in 1836, during the voyage of the *Beagle*.

In 1815 H.M.S. *Northumberland* brought Napoleon to the island where he remained until his death in 1821, the cause of which has been variously attributed to gastric carcinoma, gastric ulcer, amoebic hepatitis, typhoid fever, malaria and undulant fever. Recently it has been claimed that a sample of his hair contained toxic amounts of arsenic (Forshufvud *et al.*, 1961) and this finding provoked the question—Who did it? But the symptomatology of his final illness is inconsistent with the diagnosis of arsenic poisoning

(Turner, 1962), and there seems little reason to doubt the reliability of the original diagnosis[1] of carcinoma of the stomach which was based upon a post mortem conducted by Dr. Antommarchi in the presence of six British doctors who found a sclerosed stomach with a perforation near the pylorus.

Following the suggestion of arsenical poisoning, the *Lancet* (1961) inquired whether some of the island springs gave water with a high content of arsenic. Napoleon obtained his drinking water from the spring in Geranium Valley where he was afterwards buried, and also from the peaks whence it was conveyed to Longwood House in an aqueduct specially constructed by Sir Hudson Lowe. I sent a sample of the spring water and a sample of the aqueduct to the Government Pathological Laboratories in Cape Town, where the water was found to contain no arsenic, while the aqueduct contained only 0·15 parts per million.

The island has also imprisoned Denizula, King of the Zulus, in 1890; 5000 Boers, including General Cronjie and his wife between 1900 and 1902, and three members of the Bahreini opposition from 1959 to 1962.

Since the Suez Canal was opened in 1869, and steam replaced sail, fewer ships passed the island and even fewer needed to refuel. So began a decline in the island's importance and prosperity that has continued to this day.

GEOGRAPHY

The island is exceedingly remote from all continental land and other islands. It lies 15° 55′ south and 5° 45′ west, being approximately 1200 miles from the coast of Africa and 2000 miles from the coast of Brazil (Fig. 1.1). Ascension Island, the nearest land, lies 700 miles to the north, and Tristan da Cunha 1300 miles to the south-west.

It measures 10½ by 6½ miles, having a total area of 47 square miles, of which two-thirds are barren. The island is very singular in its appearance due to precipitous cliffs on all sides that rise over 1000 ft; as Captain Mundy (1698) put it: "It is verie rockey, hilly and steepie towards the waterside". Dominating the landscape is

[1] It is interesting to note that of the several people concerned with his final illness, the only one to make the correct diagnosis during life was Napoleon himself, and that was partly on genetical grounds.

11

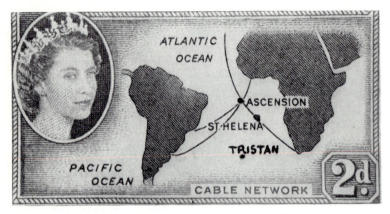

Fig. 1.1. Stamp showing the location of St. Helena.

Fig. 1.2. Jamestown, the capital and only harbour.

a central semicircular ridge of hills about 2700 ft high from which smaller ridges radiate to the coast like the pleats of a fan. The island is almost completely covered by these rugged hills and valleys leaving hardly any level ground.

The south-east trade winds blow all the year round tempering the tropical climate so that the hottest area, Jamestown, has an annual range from 57°F to 90°F, and at Varneys (1700 ft) in recent

Fig. 1.3. Country showing flax under cultivation.

Fig. 1.4. Sandy Bay. "Sir William Doveton's house, where the bold peak called Lot is seen over a dark wood of firs, the whole being backed by the red water-worn mountains of the southern shore" (Darwin, 1845).

years the shade temperature has varied between 54–88°F. Between 1957 and 1963 the annual rainfall ranged from 7·8 in. in Jamestown to 33·4 at Hutsgate near Longwood. The island soils, the result of extreme decomposition of lava, were found to be mainly acidic clays, rich in potassium, but deficient in available phosphate and magnesium (Humphrey, 1956). Thunderstorms are almost unknown, mild earthquakes have been reported occasionally in the past

13

300 years, but there has not been any volcanic activity in historic times. A geological survey in 1964 found no evidence of high background radiation and most places obtained a minimum count on the most sensitive geiger counter reading (Baker, 1964).

THE FAUNA AND FLORA

As there are several excellent and detailed descriptions of St. Helenian fauna and flora available (Mellis, 1875; Wallace, 1880; Humphrey, 1956; Loveridge, 1958–9, 1961–4; Stonehouse, 1960), the following account is merely intended as a background sketch.

The vegetation below 1500 ft is scanty and consists mainly of cactus, but above that level the island has an English or Welsh character, being covered in lantana, English broom, gorse, Port Jackson willow, poplar and Scotch pine. On the central ridge there are remnants of the original natural vegetation but the higher part of the island consists mostly of flax, pasture and forest. There are many endemic species of ferns and flowering plants (Darwin mentions 52 species) but the indigenous trees have been almost entirely destroyed by the goats which, for centuries, have been allowed to run wild, and which had already, by 1724, denuded Longwood, Deadwood and Sandy Bay.

The only crop of commercial importance is flax (*Phormium tenax*). Good quality coffee is produced in small amounts and, before the war, lily bulbs (*Lilium longiflorum*) were exported.

Many of the islanders have kitchen gardens which are generally planted with common and sweet potatoes, onions, cabbages, maize, broad and runner beans, but they seem capable of growing almost anything. There is a wide selection of fruit available, including apples, plums, paw-paws, guavas, oranges, lemons, cape gooseberries, granadillas and mangoes, but only bananas are plentiful.

There are no indigenous land mammals; those that have been introduced include cattle, sheep, pigs, goats, chickens, rabbits and two species of rat. The tree rat (the black rat of Europe) is widespread and the brown rat is relatively rare, a result of the 1924 poisoning campaign.

There are exotic species of birds of which only the wirebird is endemic. The only feral reptile is the gecko, the only amphibia are a toad and a grass frog; both introduced, the gecko probably from the Dutch East Indies, the frog from South Africa.

Widespread destruction has been caused by termites, one variety of which was introduced in timber taken from a slave ship in 1840. A new species of *Simulium* (*Eusimulium*) was discovered by Loveridge in 1962, and described by Crosskey (1965). At the same time a new genus of Simuliidae was discovered that has not yet been described. The local mosquito is *Culex pipiens*. There are a few centipedes and scorpions, but people are rarely bitten or stung by them, yet there are vast numbers of fleas that plague nearly everyone—my average score being four or five a day. Apart from annoyance, they caused little disease, although they may have been partly responsible for the widespread mild iron deficiency anaemia.

ADMINISTRATION

The Government is administered by a Governor who alone may legislate (subject to the power of Her Majesty in Council). He is advised by an appointed Executive Council and, since 1962, an elected Advisory Council. Although the Governor is usually without legal or military training, he is also the Chief Justice and Chief of the Armed Forces. For certain civil purposes, the island is divided into six districts; Jamestown, Half Tree Hollow, St. Paul's, Head o'Wain, Sandy Bay and Longwood. This division corresponds to slight cultural differences in terms of dress, eating habits and physical appearance, which the islanders recognize readily, although they are not obvious to the visitor. Ascension Island and Tristan da Cunha, which are dependencies of St. Helena, are administered locally by a person appointed by the Governor.

POPULATION

The first inhabitant, Fernando Lopez, was a Portuguese nobleman who had turned traitor in India, was captured, mutilated and granted voluntary exile accompanied by three Negro slaves. He died in 1565. The first settlers, in 1659, were Europeans who imported slaves from Malabar, Bombay, the Maldives, Calabar, the Gold Coast of Guinea, Bengal and Madagascar, the latter being particularly prized; and from 1679 onwards all ships calling at the island were required "to leave one Madagascar slave". Following the abolition of the importation of slaves in 1792, the East India Company introduced Chinese indentured labourers from their factories at Canton. In the nineteenth century Britain pursued the slavers with the same enthusiasm that she had previously applied to the trade itself, and in consequence large numbers of West African slaves were brought to the island where they were technically freed. Some were returned to Africa, most were sent to the West Indies, and an uncertain number remained on the island. Thus a century or a century and a half ago approximately one-half of the islanders were African, one-quarter were European and one-quarter Chinese. The races then mixed one with the other so that there are now very few individuals of pure descent and very few traces of Chinese or African culture or language remain.

The island's population reached its maximum at the time of Napoleon's exile, partly because of the large garrison, partly because he attracted many visitors and also because it was then an important fuelling station receiving several ships every day. The island's importance may be judged from the Governor's salary, which was £12,000 at that time. The garrison, numbering 2182, was greatly reduced after Napoleon's death and withdrawn altogether by Lord Haldane in 1906. After 1820, apart from a small number of slaves taken from captured slavers, there was virtually no immigration; and from this time on the island's prosperity dwindled, the visible consequence of which was a steady emigration that is still continuing. It is recorded that 120 persons emigrated to the Cape Colony as apprentices to farming in 1838; 280 persons went in 1872, and in 1873, 258 people went to the Cape and 442 to Natal.

TABLE 1. POPULATION CENSUS OF ST. HELENA BETWEEN 1659 AND 1966 *

Year	Population number	Remarks from records
1659		First settlers
1683	490	Excluding slaves
1714	832	Excluding slaves
1716	953	Including slaves
1718	801	Excluding garrison
1723	1110	Including garrison and blacks
1733	840	(Presumably excluding garrison and slaves)
1769	1055	Including 721 slaves
1789	1413	Including 1125 slaves. Importation of slaves prohibited in this year
1803	1563	Including 1127 slaves
1812	1732	Including 1150 slaves; also 170 Chinese labourers introduced within the last 5 years

TABLE 1 (*continued*).

Year	Population number	Remarks from records
1814	3587	European inhabitants, 736; garrison, 891; free blacks, 420; slaves, 1293; Chinese, 247
1815	3721	
1816	5511	
1817	6150	White inhabitants, exclusive civil and military, 821; slaves, 1540; free blacks, 500; company establishment, civil and military, 820; king's troops, 1475; families of king's troops, 352; Chinese, 618; Lascars, 24
1818	5468	All children born of slaves considered free from this date
1819	6061	
1820	5774	Including 1156 slaves; a garrison of 2182; Chinese, 481; free Africans, 613; and Lascars, 33
1821	5097	
1832		25 January 1832: "Slavery to be abolished as soon as possible. 869 slaves valued £37,639. Purchase money to be considered as a Loan to the Slave"
1839	4205	
1851	5490	
1861	5496	
1871	5838	Excluding garrison of approximately 500
1881	4511	
1891	3877	
1899	4270	
1910	3441	Lord Haldane withdrew garrison in 1906
1920	3718	
1930	3747	
1940	4710	
1948	4857	
1956	4642	
1966	4601	

* Taken from *Traits Relative to the Island of St. Helena*, by A. Beatson, 1816, London, W. Bulmer & Company, and from East India Company consultations and from the annual colony reports.

Slight inconsistencies of the original sources have been retained.

At present, there are 4642 St. Helenians living on the island, and probably about the same number living in South Africa and in Britain. The majority live in Jamestown, the capital, and the remainder are scattered throughout the island with a large concentration at Longwood. Over the previous years many have been employed on Ascension Island with the United States Guided

17

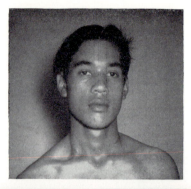

FIG. 1.5. St. Helenians.

Missile project, which employs about 100 men and a further 100–200 are employed by Cable and Wireless Ltd.

OCCUPATION

Farming is the main source of income and the largest source of employment. The main crop is flax (New Zealand hemp), which is milled and exported as fibre or tow and a small quantity as rope. The major occupations are shown below and are taken from the 1956 census.

Farming (including flaxmill workers)	322
Craftsmen	158
Other labourers	141
Relief workers	130
Management and administration	104
Police and security workers	59
Professions, teaching, etc.	44
Fishermen	24
Transport and commerce	20

The scarcity of professional fishermen is due to the lack of an efficient distribution system, and also because the country families walk to the sea once or twice a week to "fish on the rocks". Relief workers are unemployed persons who are given some financial assistance in return for light labouring for one of the government departments. The lowest permissible wage for a 5-day week is £3,[1] consequently, in comparison to England the standard of living is low, but it is much higher than in most parts of Africa. Bread, corned beef and cheese are subsidized by the Government, and many families have supplementary sources of income.

HOUSING

The majority of people live in one- or two-family houses that are built in European style. The majority are modest, one-storey white-washed cottages built of stone, or concrete breeze blocks, with corrugated iron roofs, and mud, wood or concrete floors. Most houses are well spaced and the majority have a kitchen garden of some sort.

SANITATION

All houses in Jamestown are provided with water closets, but in the country bucket latrines are generally used. Refuse is collected daily in Jamestown, but in the country it is burned, buried or dumped. Very few houses have baths or adequate washing facilities.

[1] Since writing, the minimum wage has risen to £5.

FIG. 1.6. Typical country scene. "The English or rather Welsh character of the scenery is kept up by the numerous cottages and small white houses; some buried at the bottom of the deepest valleys and others mounted on the crests of lofty hills" (Darwin, 1845).

WATER SUPPLY

The water for drinking and domestic purposes is derived from catchment areas which are found high up on the hillsides; it collects into springs whence it is directed through pipes into tanks. It is then filtered and distributed to the houses which are usually provided with indoor or outdoor taps, but in some districts it is collected in smaller water tanks and carried to the houses by hand. Inhabitants of about 190 of the most remote and scattered houses carry their water directly from springs, and many families in the country also collect rain water in butts. A few samples of spring water were sent to England for analysis in 1955, the results of which are given in Appendix 2. In most districts the water is soft, but varies from 25–225 ppm of total hardness expressed as calcium carbonate.

EDUCATION

Education is free and compulsory from 5 to 15 years of age. Until recently, although few people were illiterate, the standard was poor mainly because of the scarcity of good teachers, many of them being untrained teenagers. Most of the thirteen schools are

20

Fig. 1.7. Flax hanging out to dry.

new, spacious and pleasant, but few pupils have achieved a single pass at O-level in the G.C.E., which would not seem to be due to the lack of native intelligence or skill as far as one can judge on the basis of daily performance. For example, most men are able to farm, fish, play the guitar, cut hair, make an efficient watering can from a margarine tin, mend shoes and build their own houses.

COMMUNICATION

The only source of communication with the outside world is by sea. The Union Castle shipping line operates a passenger service which calls six times a year, and cargo vessels call with the same frequency. Within the island there are remarkably good motor roads to all districts, but many remote homes can only be reached on foot. There are now about 300 registered vehicles, but most islanders travel by foot, sometimes using a donkey to carry baggage.

HEALTH OF THE ISLANDERS

Previous publications

Because St. Helena has not had the benefit of recent volcanic eruption, there has never been any scientific expedition to the island,

21

and there are very few relevant publications to be found in the medical journals or elsewhere. The earliest is a brief description by Sir James Lancaster (1603) who noted that his crew recovered from dysentery and scurvy because of: "the very great store of very wholesome and excellent good green figs, oranges and lemons very fair".

Vice-Admiral Wittert (1608) who called on his way to Madagascar, shared this opinion:

> We found enough fine oranges, pomegranates and lemons to supply and refresh the crews of 5 or 6 ships. We also saw plenty of parsley, black mustard, sorrel and camomile, which herbs eaten in soups or in salads are very good for scurvy.

Sir Charles Bell (1836) in his case reports wrote of:

> A lady whose husband was an English clergyman at St. Helena consulted me about her child who had one leg much wasted in its growth. In conversing about the illness which preceded this affection in her little girl, she mentioned that an epidemic spread among all the children of the island about 3 to 5 years of age and her child was ill of the same fever. It was afterwards discovered that all the children who had the fever were similarly affected with a want of growth in some part of their body or limbs. This deserves to be enquired into.

This is considered to be the earliest poliomyelitis epidemic recorded (Gear, 1952), although I was unable to find reference to it in the local records.

In the nineteenth century, the medical press carried frequent mention of the island as a health resort; however, Gallowey (1909), who had been Governor, wrote a short, angry article to the *Lancet* entitled "The truth about St. Helena as a health and holiday resort", and Brousmiche (1887) reported that the Africans lived in deplorable conditions attended by high mortality with syphilis accounting for 42% of the hospital admissions; however, the infant mortality rate was lower than in Great Britain at that time.

Beriberi caused an average of two deaths a year in the last century; it became more common around 1900, and spread so rapidly in the Boer prisoner-of-war camps that it was thought to be an infection that the Boers passed on to the islanders (Casey, 1903). The local epidemiology of beriberi and its relationship to polished-rice consumption was reviewed by Wilkinson (1936). It was Wilkinson (1938) who was responsible for the only report of genetic disease; he gave an excellent description of the family with "probable haemophilia" that has now been shown to be Christmas disease, and he mentioned the possibility that the girl who died after tonsillectomy might be homozygous for the gene.

The WHO has recently financed studies on tuberculosis (Geyser, 1958) and on nutrition (Norris, 1958), although they were only published locally. Norris analysed haemoglobin levels and carried out a random larder survey. Since the islanders do not in general keep larders, it is doubtful if the results are reliable. The blue books and annual reports contain medical officers' reports which provide much useful information that is unavailable elsewhere, but there is no mention in them of any congenital or inherited disease. The only remark of genetic interest is from the administrator of Tristan da Cunha who recently wrote to the Governor saying that he believed that inbreeding was changing the island sex ratio.

EPIDEMIC DISEASES

Like any isolate, St. Helena has experienced many severe epidemics which have attacked a wide age-range of the population. The first of unknown aetiology is recorded by the Governor in a letter to the directors of the East India Company (Janisch, 1885).

> 1743. We have had abundance of mortality on this island. The inhabitants have been seized with a violent distemper very little inferior to the Plague that hath carried off abundance of them. The Slaves have been alike subject to the disorder with their masters insomuch that with great difficulty we have got people to work at the Fortifications. Nor hath your Honours Slaves escaped although there has been the greatest care taken to prevent their going to the Planters houses where this distemper raged. They are seized with a violent oppression at their stomach and pain in the small of the back and bowels attended with a strong fever and generally die in four or five days.
> Result of Postmortem on a couple of Slaves:
> 1st. Harry. Pericardium much extended with a greater quantity of water in it than usual. The right ventricle very large. On opening it I extracted three distinct pieces of flesh about an inch and a quarter in length not adhering to any part of the ventricle with a large quantity of coagulated blood. Right lobe of Lungs adhering closely to Pleura a little imposthumated.
> 2nd. Dick. The thorax full of extravasated water—the Heart larger than I ever saw—right ventricle I extracted a large brown viscous substance which almost filled the ventricle—those substances I take to be Polipuses. Right lobe of Lung entirely imposthumated. The disease chiefly amongst the Company's Slaves. What's worse of all they prove to be our very best working fellows that drop off.

Perhaps the most dramatic was the measles epidemic of 1807 which Janisch also records:

May 3rd, 1807. Measles introduced by a fleet on 24th January. The information of the existence of the disease in the fleet reaching the Governor too late to avert the most direful calamity that has ever befallen your island. The register of Burials from 1st March to 1st May exhibited a list of 102 blacks and 58 whites but many more blacks have been carried off, the exact number of whom has not yet been ascertained as they had not been christened and their burials of course not registered. The extent of distress which for a time pervaded all ranks is beyond description as in many families, including servants, every person was at the same moment incapable of the least exertion.

In the subsequent epidemics listed below, which are compiled from the annual colony reports, it is difficult to be sure of the reliability of the numbers given. Nissen (1947) reported an attack rate of 19·2 per 1000 in the poliomyelitis epidemic of 1945 but this is almost certainly an underestimate. A nurse living on the island, who knows the islanders well, informs me that the figures given in this report are erroneous because they neglect the fact that a considerable proportion of those who were affected by the disease stayed at home believing that the local remedy of daily rubs and passive movements ensured a more rapid recovery of function.

1843   "The Measles did not show themselves in the Island again until 1843 when they were brought from the Cape and produced much distress. It deserves remark that in no case in 1843 did any of the survivors of 1806 have a recurrence of Measles, but Captain Robert Wright, the only one of the surviving officers of the Garrison who was absent in South America in 1806, died from the disease immediately on its introduction in 1843".

1886   Measles. 113 cases. 8 deaths.

1887   Diphtheria.

1893   Diphtheria, whooping cough, 31 children under 10 years of age died.

1900   Influenza—"more severe than any epidemic since the 1807 epidemic of measles"—it did not attack the Boer prisoners, or the troops guarding them. Death rate = 33·3 per 1000.

1903   Whooping cough. Very few children escaped infection, although only one death.

1906   Diphtheria.

1909   Influenza epidemic.

1914   Whooping cough, influenza.

1916   Scarlet fever.

1917   Chicken pox.

1918 The world pandemic of influenza bypassed St. Helena. In 1932, Dr. Stewart Harris found a trace of swine virus antibody in very few of the 23 specimens of sera that he examined.

1924 Influenza.

1928 Whooping cough, chicken pox. Nine deaths from whooping cough between 1920 and 1929.

1932 Measles—2200 cases.

1935 Chicken pox. One death from poliomyelitis, apparently no epidemic.

1945 Poliomyelitis—11 deaths. 217 cases, an attack rate of $19 \cdot 2$ per 1000.

1948 Whooping cough—7 deaths.
      Acute nephritis—77 hospital admissions.

1951 Mumps attacked 90% of the population with a high rate of orchitis and mastitis.

1955 Whooping cough—3 deaths.

1958 Poliomyelitis—36 cases. No deaths.

1961 Over 2000 cases of a febrile illness resembling Bornholm's disease.

HEALTH SERVICES

The health department has a budget of about £20,000 annually, which is administered by the senior medical officer. The medical services are virtually free, nominal charges being made for hospital admissions, medical attention and drugs; although the senior medical officer is permitted to charge when necessary. He is assisted by a medical officer in the routine work of running the hospitals and clinics. There is a 58-bed general hospital in Jamestown and a 20-bed mental hospital in Half Tree Hollow, both of which were recently built out of the Colonial Development and Welfare Fund. There is a dental officer, an able locally trained dentist, a nursing staff of about twelve nurses with a well-qualified St. Helenian matron.

The hospital is reasonably well equipped, with X-ray facilities and a physiotherapy department and a passably modern operating theatre. The WHO has provided a small pathological laboratory equipped with a spectrometer, centrifuge and good basic apparatus. They also trained a local technician. The Population Genetics Research Unit kindly loaned a transistorized electrocardiogram which was invaluable in the country districts which are not supplied with electricity.

In public health matters, the senior medical officer is assisted by a health inspector, his assistant, a health sister and her assistant. The health sisters carry out routine school inspections, hold antenatal and postnatal clinics, and carry out the occasional home delivery, which does not amount to more than one or two per year.

HEALTH STATISTICS

All births are notified by a registered midwife, and all death certificates are issued by the government medical officers, which guarantees a certain level of accuracy of the vital statistics upon which genetic and demographic studies usually rely. It is only since 1900 that births and deaths have been regularly certified by doctors and the old records are liable to be erratic as can be seen from the following certified causes of death taken from the late nineteenth-century death registers.

Maria C.—"died of old age accelerated by senile delay".
Mary G.—"death due to labour (of mother)".
Child of Sarah W.—"from symptoms".
John G.—"from natural causes hastened by misadventure".
Albert J.—"from teeth and fever".

And here is a coroner's verdict dated 3 June 1855: "Death by visitation of God."

Finally, Robert P. died "of mortification".

It must be said, however, that more recently the standards have improved, and over the past 50 years certification would seem no less reliable than in Great Britain.

A summary of the major causes of death from 1890 to 1960 is given in Tables 2 and 3. Familiarity with the islanders' surnames enabled me to distinguish between expatriates and islanders listed in the death register; only the latter have been included in Table 3.

NUTRITION

An accurate assessment of a community's nutrition is generally difficult to achieve, perhaps more so in St. Helena where some people are wishing to conceal poverty and others a private source of income,

TABLE 2. ANALYSIS OF THE MAJOR CAUSES OF DEATH
Number of individuals dying by cause and by decade

Decade	Cancer (all types)	Pulmonary tuberculosis	Infections	Cerebrovascular accident	Bronchitis	Ischaemic heart disease	Syphilis	Infant mortality (approx. figures)	Malnutrition	Number living
1890–9	30	30	74	42	53	0	4		8	3882
1900–9	26	39	27	40	40	2	0	120	9	3342
1910–19	22	25	22	46	17	2	3	60	9	3475
1920–9	26	23	29	55	17	2	1	100	25	3747
1930–9	41	7	27	63	60	0	3	130	13	3996
1940–9	45	11	30	55	36	11	0	80	4	4728
1950–9	54	8	7	51	17	29	4	35		4642

TABLE 3. NUMBER OF DEATHS DUE TO CARCINOMA—ACCORDING TO THE DEATH REGISTER

Decade	Lung	Stomach	Colon	Oeso-phagus	Brain	Breast	Cervix uteri	Others	Total
1890–9		4			1	3	4 uterus	1 pharynx, 2 liver, 1 prostate, 2 ovary, 2 bladder, 1 palate, 1 neck	22
1900–9		10	3			1	1 uterus 2 cervix	1 tongue, 2 liver, 1 prostate, 1 ovary, 1 sarcoma, 1 heart	24
1910–19		7	2	2	1		3	2 liver, 1 penis, 1 ovary, 1 testes	20
1920–9		4	7	1		2	3	4 liver, 2 Hodgkins disease, 1 leukaemia	24
1930–9	1	15	7	3			10 cervix	1 tongue, 1 liver, 1 fib. uterus, 1 tonsil, 1 epiglottis and palate	41
1940–9		16	4	2	4	5	4	1 liver, 1 prostate, 1 ovary, 1 rectum, 1 lymphosarcoma, 1 palate, 2 bladder	43
1950–9	1	9	5	2	4	5	10	1 palate, 1 lip, 1 liver, 1 skin, 2 ovary, 1 gall bladder, 2 sarcoma, 1 vagina, 1 penis, 1 bladder	50

and where the minimum team of four recommended by Sinclair (1959) to assess nutrition was obviously impractical. It was decided that the best approach was by verbal questionnaire. A district of the island comprising 320 persons was chosen where the social and population structure was representative of the whole. Every house-holder in this district was given the questionnaire, and everyone agreed to co-operate. An area sample was chosen instead of a random sample because it was easier to organize, because the people in this district were unusually friendly, and I knew of a well-respected local farmer who agreed to visit the families personally and fill in their replies. It was natural for him to put the questions in dialect and take into account the usual manner of buying food, and we decided that it would be preferable if the replies were recorded anonymously. This approach seemed to satisfy the islanders and judging from the close correspondence of their replies to the total colony imports, it was clear that they gave accurate informa-tion. The only difference between both estimates was for sugar consumption which was due to the unusually large import in 1960 compared to other years (Appendix 3).

According to the replies to the questionnaire, the islanders consume 3335 calories per adult per day, of which $10 \cdot 5\%$ is derived from protein and $60 \cdot 5\%$ from carbohydrate, which includes a daily sugar consumption of $131 \cdot 5$ g (standard deviation $= 40 \cdot 8$ g), equivalent to $0 \cdot 29$ lb (S.D. $= 0 \cdot 09$ lb). The intake of fats is 29% of the calories, which is lower than in many Western countries and one-tenth of it is marine (unsaturated) fat.

These results, which are set out in detail in Appendix 3, are consistent with the clinical impression of a generally well-nourished community with no obvious malnutrition (for distribution of body weight about 14lbs see table 11). Vitamin deficiency disease occurred occasionally in old people living alone, but there has been no report in recent years of beriberi, rickets or scurvy, and kwashior-kor has never been recorded. In 1956 some of the school children were considered to show a mild vitamin A deficiency but there was no evidence of such a deficiency during the course of this study. Free powdered milk is supplied by UNICEF to all children up to 15 years of age and vitamins A and D capsules are given to all expectant mothers and school children, providing them with 32,000 units of vitamin A and 6400 units of vitamin D per month. Bread is the major source of calories, having superseded rice in 1939 which,

29

in its turn, slowly replaced coco-yam over the previous 100 years. Since 1957, imported flour has been fortified with vitamin B, iron and calcium, thus it is unlikely that there is a deficient intake in these substances.

The high consumption of sugar is consistent with the clinical finding of widespread dental caries and even infants were not infrequently subjected to total dental clearance. The possible relationship of sugar consumption with ischaemic heart disease is discussed in Chapter 3.

PRESENT STATE OF HEALTH

The impression obtained from the out-patient clinics is that the common types of illnesses were similar to those seen in the average British practice; instead of tropical disease and malnutrition, which might be expected of an underprivileged community within the tropics, the common ailments were diabetes, varicose ulcers, high blood pressure, bronchial asthma, cardiac failure and neuroses. This impression is confirmed from scrutiny of the causes of death obtained from the death registers from 1890 to 1960. On the whole, the list is not much different from the causes given in the Registrar-General's report for Great Britain. The most notable exception is the rarity of carcinoma of the lung which was first recorded in 1939 and has since been recorded only twice. Examination of a lady, aged 65 years, revealed a missing left mandible which she said was removed at 8 years of age by Dr. Arnold who had told her that she had a malignant tumour. It is possible that this was an example of Burkitt's lymphoma, although there are other explanations. My general impression, which is confirmed by the expectancy of life tables and the low infant and maternal mortality rates (Table 4), is that the St. Helenians are a surprisingly healthy population. Indeed, for the past 6 years the island has won a trophy awarded to that community within Britain or the Commonwealth with the best results in these fields. The following summary of disease patterns listed by systems is compiled from the death registers and personal notes made at out-patient clinics. It is intended as a guide rather than a precise indication of prevalence, although for conditions whose ascertainment was complete, such as goitre, accurate figures are given.

TABLE 4. INFANT MORTALITY IN ST. HELENA AND UNITED KINGDOM BETWEEN 1900 AND 1960

Decade	Infant mortality rates per 1000 live births	
	St. Helena	United Kingdom
1900–9	114	128
1910–19	56	100
1920–9	96	72
1930–9	95	59
1940–9	65	43
1950–9	29	25

Cardiovascular system. Degenerative cardiovascular disease is the major cause of death, and widespread atherosclerosis was frequently seen at postmortem. Ischaemic heart disease (I.H.D.) is increasing in frequency (Chapter 3), but rheumatic heart disease was infrequently given as a cause of death. Two children were seen with rheumatic fever. High blood pressure was common.

Respiratory system. Bronchial asthma was widespread particularly in children, but death due to status asthmaticus was apparently rare. Emphysema and cor pulmonale were common, particularly among flax workers, which association was first noted by Ramazzini (1700). However, X-ray examination of people who were chronically exposed to flax dust revealed no pulmonary fibrosis. As mentioned above, carcinoma of the bronchus was rare, which might be anticipated in a community that could not afford to smoke until recently, and even now men smoke only six cigarettes a day on the average (S.D. = 4), and few women smoke at all. The prevalence of pulmonary tuberculosis was low, ten cases were diagnosed between 1951 and 1957, and during the period of investigation there were only two active cases seen and one death from acute pulmonary tuberculosis. No other country surveyed by WHO has so low a prevalence of tuberculosis infection as that observed in St. Helena, and there is a low incidence of specific immunity as shown by the Mantoux test (Geyser, 1958). Almost everyone under the age of 20 years was negative to the 5 and 100 T.U. dose.

Alimentary system. Carcinoma of the oesophagus, stomach and colon appeared to be relatively common, whereas gastric and duodenal ulcers appeared to be uncommon. It is widely stated by the French that hepatitis is endemic, but I know of no evidence to

31

support this notion. During the period of investigation, the only case of hepatitis was due to leptospirosis icterohaemorrhagica.

Central nervous system. Cerebrovascular accident is one of the most frequent causes of death. Tetanus has been responsible for 2–3 deaths per decade, and the recent poliomyelitis epidemics in 1945 and 1958 have left many people with residual paralysis. Hysteria is a common neurotic manifestation (generally attributed to evil eye by the islanders), although anxiety and depressive states are seen more frequently. No one with an island name appears to have died of multiple sclerosis. Thirteen people gave a history of grand or petit mal epilepsy, nine of whom were currently undergoing treatment.

Endocrine system. There were three people with active thyrotoxicosis, and two whose thyrotoxicosis had been controlled by thyroidectomy; seven with a thyroid adenoma and one diffuse goitre (not counting two pregnant women with mild thyrotoxic goitre). There were thirty-one diabetics undergoing treatment. The disease was easy to control and there have been no deaths recorded from diabetic coma.

Locomotor system. Four people are known with rheumatoid arthritis. Gout was diagnosed in one family. Inherited abnormalities, including a unique family of dwarfs, are described in Chapter 4.

Skin. Urticaria and flax-sensitive dermatitis were the most common diseases seen. A unique blistering leg cellulitis is described in Chapter 4.

Venereal disease. Syphilis and gonorrhoea were seemingly very rare.

Violence. Deaths due to felony are uncommon, although suicide and accidental deaths probably occur about as often as in this country.

Blood disease. Leukaemias are not often recorded, but there were scanty pathological laboratory facilities before 1958. Mild iron deficiency anaemia often occurs in pregnant women and occasionally among the remainder of the population. A 2-year-old mongol child with a haemoglobin level of 21% was admitted to the hospital in cardiac failure.

Tropical disease. These diseases are uncommon. There is no malaria, filariasis, trypanosomiasis, schistosomiasis, ankylostomiasis, onchocerciasis, yaws, cholera, or smallpox. The only patient with leprosy is long-standing and inactive. The only parasitic diseases were amoebic dysentery, which is rare, and ascariasis, which is common, although it does not cause much ill health. In the old death registers hydatid disease and trypanosomiasis appear once, filariasis and malaria twice.

THE ISLAND CULTURE

The island doctor has an unrivalled opportunity to learn to understand the islanders, particularly during the first 3 months, when they come to the clinic in unusually large numbers, ostensibly to present their chronic complaints, but in reality to take a good look at the new doctor.

Inevitably, the first facet of their culture with which I became familiar was their attitude to illness and, in particular, their criteria of successful medical treatment which are very different from ours, as the following letters about a patient will show:

Man and Horse,
28th March

DEAR DOCTOR,

I asking you please if it will be convenient for you to come over a few minutes to see Ahab. He is not well again, with swollen legs and feet.

He asks for tablets to make more water because he is passing very little.

Thanking you,
ETHEL

On receiving this letter, I visited the patient who had congestive cardiac failure and treated him with digoxin and chlorothiazide. A few days later, I received the following letter concerning the same patient.

Man and Horse,
30th March

DEAR DOCTOR,

Ahab wished to thank you very much, he is feeling very much better, the tablets act like magic. The only difficulty with him now is feet and legs still swollen and feeling a little weak but stomach normal.

Thanking you,
ETHEL

The reason that the patient was very pleased, in spite of the fact that the important symptoms remained unchanged, was that the drug had had a purgative effect and this, to the islanders, was an important indication of successful treatment, to such an extent that, in its absence, the islanders will sometimes report that a drug has not worked even though the symptoms of which they were originally complaining have improved. This example shows how easy it is to draw wrong conclusions from what the islanders say, with an insufficient knowledge of their culture and, because of this problem, it was decided not to use a therapeutic trial of coronary vasodilators in the diagnosis of I.H.D.

Islanders often use and understand words in surprising ways. On asking the question "How old are you?", one is apt to receive the reply, "Very well thank you". To discover the age, the trick is to ask, "How old you are?", which the islander immediately understands and gives an accurate answer, except that he uses the age at his next birthday. On asking an islander "Do you worry?" one very often gets something like the following answer, "Yes, I worry a bit against the hill when I fetch water". Worry, to the islander, means physical discomfort on exertion rather than mental anguish, and this information is useful in getting a history of exercise tolerance, because one knows that, for this purpose, one has to ask, "Can you worry against the hill?".

On one occasion, I received a telephone call requesting a home visit to a Mr. Thomas Pugh, who lived 5 miles away, and I asked for someone to meet my car at the point where the road was closest to the house. I was met, as arranged, and I asked to be shown the way to Thomas Pugh's house. After a 2-mile walk, I was startled to find the house empty. I demanded to know where Thomas Pugh might be. "I am Thomas Pugh", the man said. I had made the mistake of omitting to ask his name on meeting him and he was far too courteous to volunteer corrective information.

The islanders would often describe their symptoms in ways which, to us, would seem bizarre, e.g. "The cold was stickery as stick to your throat". This, translated, means that the patient has a chest cold accompanied by viscid sputum which is hard to cough up. An islander would come in and announce: "I feel quite done-up-fied with sensation feelings. That kind of way I feel, doctor", or, alternatively, "I've got indigestion, even in my hands". Both these statements are merely descriptions of vague malaise and would correspond to an English patient saying, "Could you give me a tonic, Doctor?, I feel run down". It was also useful to know some of the islanders' interesting versions of English terms. A cowbuckle meant a carbuncle. An abster meant an abscess and an ulster meant an ulcer. A pain in the brains meant a headache. Roses were testicles. And one man claimed the interesting experience of being in a plaster palace for 6 months.

When I gave an old man liniment methyl salicylate to rub on his "rheumatic leg", he asked whether I recommended him to massage up or down the leg, and was obviously greatly relieved when I suggested massaging the limb downwards; which method, although I was ignorant of it at the time, induces the spirit inhabiting the limb to leave at the extremity. The belief in the power of the "evil eye" is widespread on the island. The idea that certain people with the evil eye can bewitch, injure or even kill with a glance, is extremely common, and has occurred in most cultures at one time or another. Woiciki, in his book on Polish folk-lore, has recorded the case of a man who, fearing that he was afflicted with the evil eye, blinded himself so as not to injure his children. In Naples, there were certain people called *jettatore*, who were believed to have this power, and it is recorded that when one of them walked down the street, the whole street would be cleared of people in an instant. On 17 January 1716 John Batavia, one of the British East India Company's slaves in St. Helena, was placed in irons for

> thieving and pretending he is acquainted with ye Devil and that he can bewitch any body—which terrifies all the Blacks so that two of them formerly have died by being affrighted with him. His method is to go amongst any Slaves but especially the Company's in the night time and with hard words screamed out aloud when he had put them in a consternation he stole away their victuals. The two that died he only layed his hand on their faces and repeated his hocus pocus words but they never recovered their fright. He is otherwise a good slave.

The Bible contains references to the evil eye, e.g. Deuteronomy: "If you will not hearken unto the voice of the Lord thy God, then all the following curses will fall on you—the man that is tender among you and very delicate, his eye shall be evil toward his brother and toward the wife of his bosom and toward the remnant of his children which he shall leave." The islanders would often quote these biblical passages in support of their belief. I did not immediately realize that this belief existed, since the islanders are very reticent about it as the following letter from a patient will show:

> Napoleon Avenue
> *October 9th,* 1961
>
> DEAR DR. SHINE,
> I wonder if you mind calling in my house I would like to consult you about my legs. I have suffered with Rheumatism for several years now and this lately I cannot get out to do my work. I have been to the hospital over 2 months ago but as usual the medicine didn't do me any good. No chemicals is any good for me in fact they make my pains worse. I know the reason but I don't know if you will agree with me.
>
> I remain, Sir,
> Yours respectfully,
> THEOBALD BRUTUS

35

When I first arrived on the island, I would have interpreted this letter on the basis of my Western experience of medicine. "This", I would have said "is a know all who has read the *Home Doctor* and wants to diagnose and treat his own condition." By now, I realized that the patient was merely telling me as openly as he could bring himself to do so that he was down with a case of evil eye and that Western remedies would be inappropriate. His real reason for consulting me was to obtain a certificate to remain away from work while under treatment from a local witch doctor.

It is important to know the various euphemisms that the islanders use in referring to witchcraft so that one can avoid them oneself in referring to their conditions. For instance, one day, I was examining a man with an early pulp space infection of the finger and finally told him that in my opinion there "might be something in it". The man hurriedly thanked me and left before I had time to give him any treatment. The islanders believe that witch doctors have the power of inserting foreign bodies such as pumpkins, rabbits, snails, goats, or even motor-cars, into various parts of peoples' anatomy and the islanders had several ways of alluding to this. They might say "He was not done right by", or "There's something extra", or "There's something in it". I had unwittingly suggested to the man that his finger was bewitched. Another example of the way in which the islanders would seize on the slightest hint as evidence of supernatural forces was the following incident.

Father Beecher lived in a house on Napoleon Avenue next door to Theobald Brutus, the man who wrote me the letter about his legs. One day the water supply serving both their houses developed a fault. Brutus and Beecher had a discussion as to what might be wrong and, finally, Father Beecher, with a twinkle in his eye, said, "I think there might be the very devil in it". Brutus recounted the conversation to me and, chuckling, he said, "And, of course, there was a devil in it too. We flushed it out with garlic water and it worked wonderful after that". Garlic water, incidentally, is a sovereign island remedy for witchcraft. Certain local medicines may possibly have some pharmacological basis; poplar bud extract, for instance, which is given as a tonic, is an ingredient of white pine syrup N.F.; worm wood (Absinthium) which is given for intestinal worms contains d-thujone, phelladrene and cadinene; lemon grass tea, which is given as a tonic and a carminative, contains 50% citral which is a gastric stimulant; and burnt toast is given for diarrhoea. However, most of the nostrums have a more doubtful basis; it would be surprising if parsley root elixir relieves acute urinary obstruction, a lady bird inserted into a tooth cavity relieves toothache, or extract of deadly nightshade heals infected skin lesions and removes stains from clothing. By contrast, Western remedies were far less exotic, nor did they generally produce alarming side effects, nor even diarrhoea; consequently, the islanders were often distrustful of them but were too polite to tell the doctor this and would return faithfully, week after week, for a supply of the pills they had no intention of using, and this made it difficult to assess treatment. The following conversation was overheard outside the clinic:

"I ain't no better."
"Do you take the doctor's treatment?"
"Do you take the pills the doctor give you when you are ill?"
"Yes."
"More fool you."

During my first 6 months on the island, my awareness of the islanders' beliefs and customs increased and this made it easier for me to gain the sympathy and support of the islanders when I started to do research. This was particularly helpful in eliciting pedigree information, since they are very reluctant to answer direct questions about family relationships and the required piece of information could only be captured after a complicated series of moves, as in a chess game. The following annotated conversation will illustrate the procedure. I was trying to discover the pedigree of an unmarried woman, Maud Cicero, who was an albino, and was questioning one of Maud's near relatives called Stella James.

"Who is Maud Cicero's father?"
"I wouldn't know."

This was, in fact, quite the wrong approach to use. It was much too direct, and if Stella had not been such a good friend, conversation would have stopped there. Furthermore, Stella's reply that she did not know was not to be taken literally, but was merely a hint to me that my first question would have to be rephrased. The conversation continued:

"Did you ever see anyone hanging around the house?"
"Well, Willy Waterincan was always there, in fact he did used

to sleep there, I know that because that it was made his wife leave him."

"Does she look like his?"

"I reckon she look a bit like his'n. I fancy so. She look like Marjorie more that what she look like a Cicero, because all the Ciceros are dark."

Stella's statement that Maud Cicero looked like Marjorie, who is Willy Waterincan's daughter, was a suggestion that Maud and Marjorie might have a common father but this did not necessarily mean that the common father was Willy Waterincan, since it was widely believed that Marjorie herself was illegitimate.

Then, Stella volunteered:

"They did say as Maud's children belonged to Bobby Waterincan (Willy Waterincan's son), but it wasn't him."

"How do you know?"

"I was along my house when the police came. They took him from the mill but he said he'd stand up before the muzzle of a gun but it wasn't him. He said they could hang him but it wasn't him."

"Who do the children look like?"

"Most like herself I fancy."

This was a hint suggesting there might have been incest. It must be explained here that Maud had had three children and after the first one, the family had called the police to investigate but nothing was said after the second child. The conversation went on:

"Seeing as they didn't make any fuss the second time, could it be someone close in the family?"

"Could be", said Stella.

"She never went out nowhereas after the first baby. She used to go fetch wood but not afterwards."

This was a hint from Stella that, since Maud had not left the house since having her first baby, the father of the second baby, at least, must have been living in the house. Then she added,

"People would say to Susan (Maud's guardian), 'There's many a flea is in your own blanket', but she couldn't pick the meaning of it", and Stella winked at me.

I knew where Maud lived and that there were three men living in the house. Without this bit of local knowledge, I wouldn't have been able to get much further. The next task was to find out which man it was.

"D'you reckon it could be Claud?" (Maud's uncle also living in the house).

"No, it couldn't be Claud."

"Why not?"

Stella burst out laughing and said, "No, he hasn't got the motions", meaning that the man in question did not seem to be particularly adventurous sexually. We had now eliminated Claud and, perhaps mistakenly, I did not consider Joseph Cicero, another inmate of the house, as a possible father, since he was 73 years old and had congestive cardiac failure. The other person in the house was a young boy in his twenties. He was Maud's brother.

"What about Harry?" Stella's eyes sparkled in acquiescence.

It is to be noted that throughout the course of this conversation Stella knew the acknowledged father of Maud's children perfectly well, but although, following the custom of the island, she would not tell me directly, she was willing to give me hints provided that I asked the right questions. The islanders were accustomed to approach all subjects in a roundabout, rambling way but, in addition to this, there was a slight reticence in revealing their cultural patterns to an outsider, and, needless to say, it was very important not to give the slightest impression that one disapproved of illegitimacy or even incest, which the islanders did not regard as particularly unnatural. For instance, having heard of a possible case of incest I mentioned it to an old lady who I knew well, having attended to her varicose ulcers for many months, "I hear a man out Cookham Point go with his daughter, do you think that could be true?" Whereupon she roared with laughter and said "Tell me what man don't. I wouldn't trust a man with a tom cat much less a she cat."

It was obviously important that the islanders should accept and, as far as possible, feel comfortable about the research that I was doing and at first I attempted to achieve this by explaining my purpose and questioning them in the traditional manner. I would always begin by explaining that I was doing research, but I think they soon forgot about this and fell to gossiping with me as they would among themselves in their homes, over a cup of coffee.

The elucidation of the pedigree often involved collating many conversations similar to the one that has just been described, together with other independent evidence. Although, as has already been explained, the collection of pedigree information had to be carried out in an informal way, the data thus collected were stringently interpreted and a pedigree was not finally accepted until it had been confirmed by many independent reports. In fact, data collected in this way give more reliable results than those which are based on a single formal interview with one member of a family which is the method preferred in this country. A subject is liable to conceal illegitimate births in the course of a single, formal interview with a comparative stranger, and it is noted that the illegitimacy rate to be observed in the pedigrees recorded by genetic studies is very much lower than that occurring in the population.

The importance of knowing nicknames in establishing pedigrees has already been implied, but this knowledge was also useful in another way, since a nickname often gave a clue to a congenital disorder, as the following examples will show:

Benny "Scissors" had gross familial genu valgum.
"Clock Eyes" Eileen had congenital nystagmus.
Sarah "noneck" had Klippel–Feil syndrome.
Ebby "Boom bang" was an albino who got his nickname in a curious way. The children, noticing his pink rabbit-like eyes, would chase him down the street, singing "Run, rabbit run," . . . and when they got to ". . . here comes a farmer with his gun", they would shout "Boom bang" in unison.

The description of the islanders that has been given is a composite one, and by no means reveals the wide range of attainments and behaviour to be found among them. Some islanders were Western in their outlook and language, but it would be very easy to underestimate the intelligence of those islanders who had not yet acquired Western ways. For instance, the following letter, at a first superficial glance, gives the impression of an unintelligent man, but a closer inspection reveals a beautifully observed, circumstantial and logical account of a prolonged illness.

Dec. 7, 1960

DEAR DOCTOR,
    I had a wound on my leg and I went up to doctor strang and he told me it was ulcers and a few weeks later he take two slides one from the skin and one from the nasel and he told me I was a Leper and I had to be islated away at Ruperts for 16 months and the treaterment that he gave me was chaulmoogric oil and then he take another slide and say it was nackertif and then I removed from there to the place weare I are now and then we

were under doctor Jafferys and he take no slides and then under doctor
Duncan and he took one slide and it was poserlived and then under doctor
Lee he take no slides and then under doctor gurd and he take no slides
and it all included to six years on the 12 December I was Islated away from
home and also doctor Cawfield take two slides one from the skin and one
from the nasel and say it was nackertif and the treaterment that I am on is
suphetrone.

ALBERT HUMPHREYS

PRACTICAL PROBLEMS

The doctor who wishes to investigate an underdeveloped
community must be prepared to face a lack of many technical
facilities which he has previously taken for granted. For example,
I was investigating a patient with leg cellulitis and realized that
I would need sheep's red cells for a Paul–Bunnell test. Sheep's red
cells were not a part of our standard equipment, but it occurred
to me that the island did have sheep. So, as the cookery recipes say,
I first found my sheep. Having bled it, I washed the cells and the
test worked perfectly.

Again, a distinguished haematologist suggested investigating the
cause of the haemorrhagic diathesis which I found on the island.
He recommended a thromboplastin generation test and a one-stage
prothrombin time, for which it is necessary to have brain extract.
This did not seem to present any problem to either of us, but when
I returned to the island I wondered where I could get a specimen
of brain. I gathered from the mortuary attendant that the islanders
were reluctant to permit the removal of material at post mortem
and, furthermore, a post mortem never coincided with episodes
of the haemorrhagic illness. In the end I was unable to do the test.

There was an occasion when I wanted to perform a salicylamide
tolerance test on the patients with hyperbilirubinaemia. I was
advised by Dr. Barbara Billing that the method I should use was
that of Fishman and Green which requires one-fifth normal sulphuric
acid. So I ordered concentrated analytical sulphuric acid from
Cape Town thinking that I would be able to dilute it. It arrived
$2\frac{1}{2}$ months later. When we diluted the acid to test normal controls
we obtained erratic results. On re-reading the instructions I noticed
that slight impurities can interfere with this test, so I checked all
previous stages of the preparation and noticed that the distilled
water was impure because of an imperfection in the still. The

41

tolerance test involves ingesting 4 g of salicylamide and since the patient was perfectly well as far as he was concerned, he was understandably reluctant to take it. In fact, only one of the patients with this condition agreed to take it at all and he, I realized, would probably only agree to take it once. I therefore had to be certain of my method before I embarked on it. I did not want to attempt anything with the sulphuric acid I had, since, apart from the fact that I was uncertain about the distilled water, the method of standardizing the strength of the acid involved weighing substances on a balance of doubtful accuracy. So, I ordered some one-fifth normal sulphuric acid direct from Cape Town. The firm sent a cable in reply explaining that they did not supply one-fifth normal sulphuric acid but indicated that they could send their product to the public analyst who would dilute it to the required strength. I sent a reply indicating that I wished them to do this, and 6 weeks later the acid arrived, but since this was not a regular product it arrived in a bottle with a cork rather than a glass stopper and the acid had reacted with the cork making it useless. I sent another cable to the firm indicating that the acid must be sent in a glass-stoppered bottle and, finally, 6 months after my first request, the reagent arrived and, happily, the test then worked perfectly.

Then there was the affair of the mice. I wanted to ascertain whether the leg cellulitis might be due to a virus and for this purpose I secured, from Johannesburg, a special breed of mice which is suitable for culturing the virus. After some difficulty with freight and quarantine regulations I managed to get them to the island. It may seem a simple matter to maintain a colony of mice on an island and so, indeed, the man who had supplied them to me made it appear. However, I was soon to be disillusioned. Since only newborn mice are suitable for inoculation with the virus, I had to arrange the size of the colony so that there was a high probability that a new litter would be produced every 1 or 2 days. This meant that I had to allow the colony to expand to a size that was too large to be contained in the cages I had brought with me. With the assistance of the St. Helenians, who are very good at these things, we managed to make thirty cages out of some biscuit tins, but my problems were by no means over. The cages had to be accompanied by feeding bottles and I had not brought enough with me. To construct a feeding bottle may appear a simple matter, but our experience was that the bore size of the outlet-tube is so critical that either the mouse could obtain no water at all or was in imminent danger of being drowned. Even this was not the end of the problem for, although I had been told that the mice were not very particular feeders, it soon became apparent that they were not very keen on the food available on the island and, in fact, preferred to eat each other. I hastily ordered some appropriate food from Cape Town, but it was 6 weeks before it would arrive and meanwhile the mice were in imminent danger either of drowning or of cannibalism. Finally, when all these problems had been overcome, and I had at my disposal a placid, well-fed colony of mice, the disease that I had originally wished to study disappeared from the island. Eventually, just before my departure from the island, one more case occurred and some inoculations were carried out.

Even such an apparently simple matter as determining the height of the population presented difficult problems. Sir George Pickering had recommended doing this, perhaps visualizing patients filing into a clinic equipped with a weight and height machine. In fact, however, some of the patients that I wished to study were unwilling to come to the clinic, and since some of them lived in inaccessible places, all the equipment that was needed had to be carted for long distances over hills and valleys to their houses. For this purpose, I acquired an alpine mountain sack. In this I placed all the standard medical equipment plus a weighing machine. In one hand I carried an electrocardiogram, and in the other my photographic equipment, leaving no room to carry the graduated 6 ft pole which would have been necessary to carry out Professor Pickering's suggestion.

This has been an account of the difficulties that faced one investigator in one underdeveloped community and it is obvious that other communities will present different problems requiring different solutions. However, it is clear, that future investigators who visit primitive communities will encounter some problems that are similar to those that were described above. It would seem, therefore, that any future research team might adopt the strategy of sending out one member, well in advance, to live in the community for 3–6 months for the purpose of getting to know the people and making a preliminary survey of the scientific problems. Then he would be able to return and advise the team on what expert assistance to seek and what equipment to bring before embarking on the main investigation.

It was, for example, of tremendous assistance to me to return to England after 16 months on the island to consult Professor Rosenheim, who gave expert advice and was able to guide me to the people who could give me the specialized assistance that I needed.

CHAPTER 2

THE AETIOLOGY OF HALLUX VALGUS

It is surprising that while mankind in all ages have bestowed the greatest attention upon the feet of horses, mules, oxen, and other animals of burden and draught, they have entirely neglected those of their own species, abandoning them to the ignorance of workmen, who, in general, can only make a shoe upon routine principles and according to the absurdities of fashion or the depraved taste of the day. Thus, from our earliest infancy shoes, as at present worn, serve but to deform the toes and cover the feet with corns, which will not only render walking painful but in some cases absolutely impossible.

(P. CAMPER, 1781)

MOST people, not having Camper's simple vision, have put forward cumbersome and contradictory theories to explain the formation of hallux valgus, which has now become the most common orthopaedic deformity seen in Great Britain (*British Medical Journal*, 1952). Thus, the deformity has been attributed to a long first metatarsal (Mayo, 1908), to a short first metatarsal (Morton, 1935), to a metatarsus primus varus (Truslow, 1925; Kleinberg, 1932; Hawkins *et al.*, 1945), to muscular malfunction (McBride, 1935; Lake, 1952), to variation in the action of tibialis posterior (Kaplan, 1955), to bow-string effect of the extensor hallucis longus (Emslie, 1939), to heredity (Truslow, 1925; McElveny, 1944) and lately, to wearing shoes (Craigmile, 1953; Barnicot & Hardy, 1955).

Currently there are two favoured hypotheses; it is said that the disease is hereditary due to some structural defect of the foot, or alternatively that it is environmental due to inadequate shoes. The following data are cited in support of the first hypothesis:

(1) Hallux valgus may be present at birth (Hoffman, 1936; Meyerding & Upshaw, 1947; McCormick & Blount, 1949).
(2) Hallux valgus has been reported to occur through several generations (Sandelin, 1922; Keizer, 1950; Johnston, 1956).
(3) There is a different sex incidence at an age when both sexes wear similar shoes (Hardy & Clapham, 1951).

45

(4) Hallux valgus does occur occasionally among unshod populations (Barnicot & Hardy, 1955; Kalcev, 1963).

Support for the second hypothesis rests on:

(1) The high frequency of hallux valgus in shoe-wearing communities (Craigmile, 1953).
(2) The low frequency of hallux valgus in barefoot communities (Barnicot & Hardy, 1955; Kalcev, 1963; James, 1939).
(3) The regression of the abnormality after wearing well-fitted shoes (Knowles, 1953; Craigmile, 1953; Kieser-Nielsen, 1953; Burry, 1957; Barnett, 1962).

It is, of course, probable that several factors could, in certain circumstances, cause hallux valgus; but it seems reasonable to inquire whether, if other things are equal, it is more likely to develop in people who wear shoes. The decision to resolve this question in St. Helena was taken because of the unique opportunity that the island presented to detect and quantify both the deformity and the environmental factor to which it is attributed.

THE STYLES AND SOURCES OF SHOES

The most common style of shoe worn by either sex was a flat-heeled, round-toed, laced shoe; although since 1960 there has been a tendency for women to wear high heels with narrow toes and for men to wear open sandals with a thong between the first and second toe. Shoes were purchased by mail order from South Africa and Britain or from the local stores; neither of which gave a choice of fitting.

By tradition the islanders did not wear shoes but they have gradually acquired the habit for various reasons over the past 50 years. The tradition persists for young children, few of whom wear shoes before the age of 5, after which the majority go unshod until they leave school at 15. Thereafter, most people sooner or later acquire the habit, sometimes as a luxury on retirement. If asked the reason why they started, they replied that it was required of teachers, government clerks, or domestic servants, or necessary for the protection of their feet, or good for their health as it was widely believed that shoes were a prophylactic against ill health, "my mother put me in shoe leather from childhood up because I used to be very sickly".

METHOD OF INVESTIGATION

A. *Separation of the shod from the unshod*

Most people who were habitually barefoot were polite enough to wear shoes when attending for medical treatment, thereby making it a little more difficult to separate the two groups. However, the splayed, engrained hyperkeratotic appearance of the habitually barefoot is so striking that it was hardly necessary to ask people whether they usually wore shoes. If they did, they were asked when they started; which was the sort of event that they were able to recall accurately. When, on occasions the reply was hesitant, I would check by asking neighbours, "how long has Mr. T. been in shoe leather?", and I was pleased to discover that both estimates invariably corresponded.

B. *Measurement of hallux valgus*

While the patient stood with the heels and the big toes touching, the feet were inspected, and the angle separating the big toes was recorded to the nearest 10 degrees. If the toes were separated by less than 10 degrees (5 degrees each foot) hallux valgus was considered to be absent because the big toe is capable of at least 5 degrees of abduction.

After examining half the population, Dr. Catherine Hollman suggested that a more precise measure of the angle of hallux deviation could be obtained using a foot board produced by the Shoe and Allied Trades Research Association. Not wishing to repeat all the previous examinations unnecessarily, a 10% random sample was chosen to see whether the foot board gave different results from the visual method, but it did not (Table 5). There were differences in estimating slight hallux valgus but overall the results of both methods were very similar, the correlation coefficient was $+ 0\cdot86 \pm 0\cdot0087$.

In order to maintain a uniform standard, visual assessment, which was quicker to make and evidently no less reliable, was used for the remainder of the population.

RESULTS

Whichever way the findings were analysed, hallux valgus was strongly associated with shoe-wearing. The population was divided into groups according to the number of years that shoes had been

47

TABLE 5. COMPARISON BETWEEN VISUAL AND FOOT BOARD GRADING OF HALLUX VALGUS ON RANDOM SAMPLE OF 444 PEOPLE (FEET SCORED SEPARATELY)

	Hallux angle (in degrees)	Visual assessment				Total
		0°–5°	6°–15°	16°–25°	26°+	
FOOT BOARD ASSESSMENT	0°–5°	570	40	—	—	610
	6°–15°	96	85	33	—	214
	16°–25°	2	12	26	7	47
	26°+	—	—	7	10	17
TOTAL		668	137	66	17	888

worn (Table 6). The mean angle of hallux deviation in each group was calculated (Table 7) and in case the inclusion of lesser degrees of hallux valgus introduced bias, the proportion of each group with severe hallux valgus (more than 15 degrees) was calculated (Fig. 2).

Subdividing the sample by age weakened the effect of shoe-wearing slightly, but at every age the amount of hallux valgus increased if shoes were worn (Table 8). In order to nullify the influence of other factors, a 10% random sample of the barefoot was matched with shoe-wearers having the same age (decade), sex, class, marital status, occupational type, body weight ($\pm$ 10 lb) and physical activity (walk the same number of miles per day $\pm$ 2). A comparison of these matched pairs showed slightly less than a twofold increase in incidence of hallux valgus in the shoe-wearing group, which was a significant difference but smaller than anticipated from the results shown in Fig. 2. On Dr. N. J. Bailey's advice the effect of all concomitant variables was estimated by multiple regression analysis using the programme S101 on Oxford University Biometry Department's Elliot 803 computer (Table 9). As the

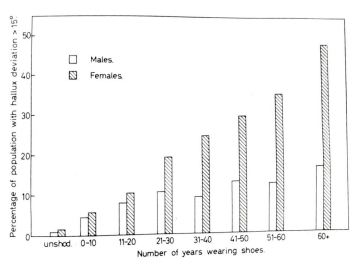

FIG. 2.1. Proportion of the population, by years, in shoes showing hallux valgus deviation greater than 15 degrees.

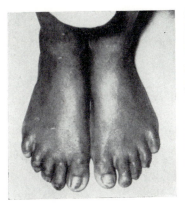

FIG. 2.2. Usual appearance of the habitually barefoot.

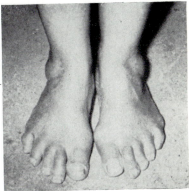

FIG. 2.3. Hallux valgus in the habitually barefoot.

programme rejected individuals with missing data for any variable, a second data tape was prepared omitting those columns that gave a non-significant t-value on the first run, thereby increasing the number of degrees of freedom.

49

TABLE 6. SHOWING NUMBERS OF FEET WITH/WITHOUT HALLUX VALGUS OF VARYING SEVERITY TABULATED ACCORDING TO DECADE AND NUMBER OF YEARS SHOE-WEARING: DATA FROM 1897 MALES

DURATION OF SHOE-WEARING	Age of individuals in years																			
	0-9 years					10-19 years					20-29 years					30-39 years				
	0°	6°-15°	16°-25°	26°+	Total	0°	6°-15°	16°-25°	26°+	Total	0°	6°-15°	16°-25°	26°+	Total	0°	6°-15°	16°-25°	26°+	Total
0	506	0	0	0	506	884	33	11	0	928	222	30	2	0	254	118	24	2	0	144
-2	0	0	0	0	0	55	12	2	1	70	43	3	2	0	48	12	4	4	0	20
-4	56	0	0	0	56	26	8	2	0	36	51	16	5	0	72	16	5	1	0	22
-6	0	0	0	0	0	88	2	2	0	92	52	13	3	0	68	20	1	1	0	22
-8	0	0	0	0	0	4	0	2	0	6	29	7	0	0	36	5	1	0	0	6
-10	0	0	0	0	0	16	4	2	0	22	36	4	4	0	44	20	2	0	0	22
-15	0	0	0	0	0	6	0	0	0	6	14	0	0	0	14	20	6	0	0	26
-20	0	0	0	0	0	0	0	0	0	0	5	1	2	0	8	28	14	2	0	42
-25	0	0	0	0	0	0	0	0	0	0	0	0	0	0	0	6	0	0	0	8
-30	0	0	0	0	0	0	0	0	0	0	0	0	0	0	0	10	5	2	3	18
30+	0	0	0	0	0	0	0	0	0	0	0	0	0	0	0	0	0	2	0	2
TOTAL	562	0	0	0	562	1079	59	21	1	1160	452	74	18	0	544	255	62	12	3	332

TABLE 6 (continued). Showing Numbers of Feet With/Without Hallux Valgus of Varying Severity Tabulated According to Decade and Number of Years Shoe-wearing: Data from 1897 Males

Age of individuals in years

DURATION OF SHOE-WEARING	40-49 years					50-59 years					60 + years					Totals				
	0°	6°-15°	16°-25°	26°+	Total	0°	6°-15°	16°-25°	26°+	Total	0°	6°-15°	16°-25°	26°+	Total	0°	6°-15°	16°-25°	26°+	Total
0	116	18	0	4	138	138	14	4	0	156	123	16	7	4	150	2107	135	26	8	2276
—2	17	9	0	0	26	12	2	2	0	16	10	0	0	0	10	149	30	10	1	190
—4	16	0	2	0	18	11	7	2	0	20	0	1	1	0	2	176	37	13	0	226
—6	8	0	0	0	8	14	0	0	0	14	6	2	0	0	8	188	18	6	0	212
—8	6	2	0	0	8	6	2	0	0	8	1	1	0	0	2	51	13	2	0	66
—10	6	2	2	0	10	16	0	0	0	16	15	3	0	0	18	109	15	8	0	132
—15	69	3	2	0	74	4	0	2	0	6	10	2	3	1	16	123	11	7	1	142
—20	41	15	9	1	66	17	12	3	0	32	16	0	0	0	16	107	42	14	1	164
—25	12	4	0	0	16	4	0	0	0	4	2	4	0	0	6	24	8	2	0	34
—30	14	15	3	0	32	17	5	4	0	26	13	1	4	0	18	54	26	11	3	94
30 +	26	0	4	0	30	81	18	3	2	104	60	36	20	6	122	167	54	29	8	258
TOTAL	331	68	22	5	426	320	60	20	2	402	256	66	35	11	368	3255	389	128	22	3794

51

TABLE 6 (continued). SHOWING NUMBERS OF FEET WITH/WITHOUT HALLUX VALGUS OF VARYING SEVERITY TABULATED ACCORDING TO DECADE AND NUMBER OF YEARS SHOE-WEARING: DATA FROM 1666 FEMALES

DURATION OF SHOE-WEARING	0–9 years					10–19 years					20–29 years					30–39 years				
	0°	6°–15°	16°–25°	26°+	Total	0°	6°–15°	16°–25°	26°+	Total	0°	6°–15°	16°–25°	26°+	Total	0°	6°–15°	16°–25°	26°+	Total
0	482	0	0	0	482	559	37	12	0	608	98	12	6	0	116	121	12	1	0	134
—2	0	0	0	0	0	51	21	2	0	74	23	11	4	0	38	14	5	3	0	22
—4	68	0	0	0	68	35	16	3	0	54	17	13	4	0	34	14	3	1	0	18
—6	0	0	0	0	0	101	21	4	0	126	30	13	1	0	44	10	4	0	0	14
—8	0	0	0	0	0	6	4	0	2	12	18	7	5	0	30	10	3	1	0	14
—10	0	0	0	0	0	21	9	6	0	36	26	16	0	0	42	10	8	2	0	20
—15	0	0	0	0	0	10	11	3	0	24	24	10	2	0	36	26	18	0	0	44
—20	0	0	0	0	0	0	0	0	0	0	18	10	8	0	36	39	5	6	0	50
—25	0	0	0	0	0	0	0	0	0	0	5	7	0	0	12	20	0	4	0	24
—30	0	0	0	0	0	0	0	0	0	0	0	0	0	0	0	20	27	6	1	54
30 +	0	0	0	0	0	0	0	0	0	0	0	2	0	0	2	0	1	3	0	4
TOTAL	550	0	0	0	550	783	119	30	2	934	259	101	30	0	390	284	86	27	1	398

Age of individuals in years

TABLE 6 (continued). SHOWING NUMBERS OF FEET WITH/WITHOUT HALLUX VALGUS OF VARYING SEVERITY TABULATED ACCORDING TO DECADE AND NUMBER OF YEARS SHOE-WEARING: DATA FROM 1666 FEMALES

Age of individuals in years

DURATION OF SHOE-WEARING	40–49 years					50–59 years					60 + years					Totals				
	0°	6°–15°	16°–25°	26°+	Total	0°	6°–15°	16°–25°	26°+	Total	0°	6°–15°	16°–25°	26°+	Total	0°	6°–15°	16°–25°	26°+	Total
0	47	1	1	1	50	56	6	0	2	64	45	12	3	0	60	1408	80	23	3	1514
– 2	14	2	0	0	16	2	4	2	0	8	6	2	0	0	8	110	45	11	0	166
– 4	7	9	4	0	20	14	2	2	0	18	6	0	0	0	6	161	43	14	0	218
– 6	14	4	0	2	20	2	0	0	0	2	0	0	0	0	0	157	42	5	2	206
– 8	4	2	0	0	6	4	5	1	0	10	1	1	0	0	2	43	22	7	2	74
– 10	8	4	0	0	12	11	3	2	0	16	5	1	2	0	8	81	41	12	0	134
– 15	2	2	4	0	8	0	0	0	0	0	3	1	2	0	6	65	42	11	0	118
– 20	23	3	4	0	30	2	2	6	0	10	8	2	4	2	16	90	22	28	2	142
– 25	17	4	3	0	24	1	1	0	0	2	2	2	2	2	8	45	14	9	2	70
– 30	19	25	13	5	62	26	19	17	2	64	12	5	3	2	22	77	76	39	10	202
30 +	35	24	12	3	74	47	38	22	9	116	110	69	80	33	292	192	134	117	45	488
TOTAL	190	80	41	11	322	165	80	52	13	310	198	95	96	39	428	2429	561	276	66	3332

TABLE 7. MEAN HALLUX VALGUS ANGLE (IN DEGREES) FOR THE BAREFOOT AND VARIOUS GROUPS OF SHOE-WEARERS: DATA FROM 1852 MALES AND 1663 FEMALES

	Males			Females		
Years shoes worn	No. of people	Mean hallux angle	S.D. of mean	No. of people	Mean hallux angle	S.D. of mean
Barefoot	1141	0·8°	3·3°	760	0·9°	3·6°
< 10	338	2·6°	5·4°	353	3·6°	6·3°
10 −	111	3·3°	6·2°	133	5·5°	6·8°
20 −	94	4·7°	7·4°	110	6·2°	8·3°
30 −	66	4·1°	6·6°	113	9·3°	8·4°
40 −	46	4·6°	8·4°	78	9·6°	9·5°
50 −	32	6·2°	8·7°	60	11·2°	9·8°
60 +	24	6·2°	7·7°	56	13·4°	10·8°

TABLE 8. MEAN HALLUX VALGUS ANGLE (IN DEGREES) BY AGE-GROUP AND SEX FOR BAREFOOT AND SHOE-WEARERS. DATA FROM 3006 ISLANDERS OVER 10 YEARS OLD

MALES

	Barefoot			Shoe-wearers					
				1–8 years			1–60 years		
Age-group (Years)	No. of feet	Mean angle (degrees)	S.D.	No. of feet	Mean angle (degrees)	S.D.	No. of feet	Mean angle (degrees)	S.D.
10 −	928	0·6	2·8	204	2·0	5·2	232	2·1	11·6
20 −	254	1·3	3·6	224	2·6	5·3	290	2·6	5·5
30 −	144	1·9	4·3	70	3·3	6·3	188	3·6	6·6
40 −	138	2·2	5·9	60	2·5	5·1	288	3·4	6·3
50 −	156	1·4	4·2	58	3·3	6·0	246	3·4	6·5
60 +	150	2·7	6·8	22	2·7	5·5	217	5·8	8·4

FEMALES

	Barefoot			Shoe-wearers					
				1–8 years			1–60 years		
Age-group (Years)	No. of feet	Mean angle (degrees)	S.D.	No. of feet	Mean angle (degrees)	S.D.	No. of feet	Mean angle (degrees)	S.D.
10 −	606	1·0	3·6	266	3·2	5·8	326	3·8	6·2
20 −	116	2·1	5·2	146	4·9	6·7	274	5·0	6·5
30 −	134	1·0	3·3	68	3·7	6·2	264	4·9	6·9
40 −	50	1·2	5·2	62	5·0	7·6	272	6·9	8·5
50 −	64	1·9	5·9	38	5·5	7·2	246	8·6	9·0
60 +	60	3·0	5·6	16	1·9	4·0	368	10·5	10·5

TABLE 9. ESTIMATES OF SOCIOLOGICAL EFFECTS ON THE DEVELOPMENT OF HALLUX VALGUS

Independent sociological variable	Males (total 1001)		Females (total 920)	
	Partial regression coefficient (b)	Significance (t)	Partial regression coefficient (b)	Significance (t)
Years shoe-wearing	0·017	5·64 ***	0·043	7·44 ***
No. of children	0·012	0·85	− 0·037	1·63
Age	0·002	0·94	0·010	2·05 *
Body weight	0·002	1·24	− 0·003	2·89 **
Occupation	—	—	—	—
Physical activity	—	—	—	—
Smoking	—	—	—	—
Race	—	—	—	—
Ischaemic heart disease	—	—	—	—
Systolic blood pressure	—	—	—	—
Diastolic blood pressure	—	—	—	—
Arcus senilis	—	—	—	—
District of residence	—	—	—	—
Social class	—	—	—	—
Child deaths	—	—	—	—

		Males		Females
Regression M.S.		$0·1251 \times 10^4$		$0·1394 \times 10^4$
d.f.	1000		919	
Residual M.S.		$0·1188 \times 10^4$		$0·1394 \times 10^4$
d.f.	986		905	

— Not significant in stepwise test.
* $P < 0·05$.
** $P < 0·01$.
*** $P < 0·001$.

DISCUSSION AND CONCLUSION

It clearly emerges from this study that the barefoot St. Helenians rarely develop hallux valgus whereas it occurs among the shoe-wearers as often as in England. Regression analysis shows a powerful association between duration of wearing shoes and hallux valgus which is not due to mutual dependence of the independent variables, since the association persists after the effect of the others has been eliminated. The large residual variance suggests that there are other unanalysed factors that contribute to the deformity; heredity is presumably one, and random variation, accentuated by the crude method of measurement, presumably another. The only reasonable explanation of this association, which also shows a biological type of time–effect relationship, is that hallux valgus is caused by wearing shoes, although it is theoretically possible that the presence of hallux valgus might cause shoes to be worn, and it is also possible that there may be a third and unknown factor causing both. Both alternatives sound unreasonable, and are unsupported by any evidence, whereas it has been demonstrated that most shoes constrict the toes (Haines & McDougall, 1954) and the aristocratic Chinese have demonstrated how effectively constricting forces may deform the feet.

Whether shoes produce their effect by direct mechanical pressure on the bone, rather as muscles produce insertion ridges, or secondarily through restricting the blood supply to muscles, or in some other way, is outside the scope of this inquiry. Nor has any attempt been made to see whether shoes confer any compensating advantages as the islanders believed. For example, it would seem reasonable to expect the shoe-wearers to show a lower incidence of leptospirosis, tetanus and foot injury.

These data show three further significant results. Firstly, there is no age of particular sensitivity to the effects of shoes; secondly, the presence of metatarsus primus varus does not determine the development of hallux valgus; and thirdly, there was no sex difference in the incidence of hallux valgus among the barefoot as there was among the shoe-wearers.

It has generally been assumed that a growing foot is more easily distorted than an adult foot, and on this assumption, at a recent conference in London, the Minister of Health was urged to control footwear in schools (*British Medical Journal*, 1965). It is surprising that this assumption should be untrue, but it clearly emerges from Table 8 that 5–10 years in shoes will cause an approximately equal amount of deformity in the previously unshod foot whether shoes are assumed during adolescence or in retirement, or at any age. It follows, therefore, that metatarsus primus varus cannot be a relevant aetiological force since it is more commonly found among older people who have never worn shoes; yet when they do, they develop hallux valgus no more frequently than young children, who infrequently have metatarsus primus varus.

It also emerges from Table 7 that wearing shoes for a given time has a greater effect on women than it does on men. Whether this is because men are less ready to assume ill-fitting shoes, or whether it is an example of sex limitation, cannot be answered. However, it cannot be sex limitation alone because no sex difference is seen in the unshod foot. It is unlikely to be a reflection of exercise, occupation or weight difference; otherwise, these factors would be expected to show a significant partial regression coefficient.

Some families seemed particularly susceptible to the effects of shoes and others rarely developed deformity regardless of their shoe-wearing habits. It is hoped to investigate the exact genetic contribution at a later date, using record linkage techniques to bring together all case records into family files; however, this is a theoretical problem because, unlike shoes, the genotype cannot be changed.

The present study corroborates many previous ones, which have found hallux valgus to be common among the shoe-wearers and uncommon among the barefoot. Engle and Morton (1931) found no hallux valgus among the barefoot Congolese, and James (1939) found a straight inner border of the foot in the Solomon islanders. The first investigators to use field measurements were Barnicot and Hardy (1955), who measured the feet of 652 Nigerians with a pedograph and compared them with 134 local Europeans; they found the mean angle in the barefoot men and women was approximately zero, although about 3% had a deviation greater than 15 degrees, and the mean angle in the European men and women was 6·9 degrees and 11 degrees respectively. Kalcev (1963) measured the feet of 4090 barefoot Malagasy who had about the same frequency of hallux valgus as the Nigerians, and he also demonstrated that the angle increases with age from − 0·64 degrees at 6 years to + 2·52 degrees at 15 years, remaining constant thereafter. Two studies have reported hallux valgus to be common among the barefoot. Wells (1931) examined a small sample of feet, both living and post mortem, in South Africa and he found that the Bantu and Bushman more commonly had hallux valgus than the Europeans. In the absence of any measurements it is likely that he was not referring to hallux deviation from the sagital plane, but rather from the long axis of the first metatarsal. More recently, MacLennan (1966) reported hallux valgus among the natives of New Guinea, which, he claimed, conflicted with a previous report by Shine (1965); but this result is due to an idiosyncratic method of estimating hallux deviation and disappears once a common standard is applied (Shine, 1966).

Other partially barefoot, yet racially homogeneous populations that have been studied, have not been found to show any abnormality until they assume shoes. Hoffman (1905) showed this to be so among Filipinos; Freiberg and Schroeder (1903) and Burry (1957) have shown a higher incidence of hallux valgus among American Negroes than African Negroes, Emslie (1939) showed hallux deviation of the terminal phalanx only of those London toddlers who wore shoes, and Lam Sim-Fook and Hodgson (1958), who examined the feet of 200 Chinese in Hong Kong, found hallux valgus among $1 \cdot 9\%$ of the barefoot compared to 30% among the shoe-wearers. Those who have doubted the implications of these studies have alleged that the comparison was made between groups with marked occupational, or socio-economic differences, which criticism does not apply to the present study where any differences that were present were allowed for in the regression analysis.

CHAPTER 3

ISCHAEMIC HEART DISEASE

Le coeur a ces raisons que la raison ne connait point.

(BLAISE PASCAL, 1670)

He [Heberden] had the good sense not to say much about the cause of the disease [angina pectoris].

(SIR WILLIAM OSLER, 1910)

AFTER 3 months in practice on the island I wondered whether the St. Helenians were immune to ischaemic heart disease (I.H.D.), for no one had complained of angina pectoris, no one had asked for fresh supplies of glyceryl trinitrate and out of nine deaths not one had been ascribed to coronary thrombosis. A community that is immune to a disease that elsewhere accounts for as many as 50% of all deaths (Oliver & Stuart-Harris, 1965) is obviously valuable, particularly as the dietary, social and clinical data that were being collected might be expected to provide an explanation. Accordingly, when patients were seen for routine examination, cardiac function was assessed at the same time.

Definition. The term ischaemic heart disease is used to mean obstructive coronary atheroma, causing angina pectoris, coronary thrombosis, or myocardial infarct.

METHOD

Apart from the well-known difficulties in detecting I.H.D. (Bronte-Stewart & Pickering, 1959) St. Helena presented its own peculiar problems; not only were symptoms described in an unusual manner but they were milder than anticipated. At first, people were given the questionnaire designed by Dr. Miall to detect I.H.D. in Jamaica, but the St. Helenians were not good at answering unfamiliar questions succinctly. When asked, "Have you ever had any pain or discomfort in your chest?" (Dr. Miall's first question) the reply might be "My ches' did used to get quite tightfied".

59

"How do you mean tightfied?"

"You know quite uselessfied, with sensation feelings, that kind of way it was doctor."

That type of reply generally required lengthy discussion to clarify the symptoms and did not encourage me to ask the remaining forty-four questions. After some preliminary mistakes it was found that the most effective discriminating question was, "Can you travel up Ladder Hill without stopping?", which invariably received the answer yes or no. The reason why people gave an immediate and reliable answer to this question was because they had walked up Ladder Hill for years, and they knew exactly how long it took them and where they usually stopped. Anyone who was able to climb Ladder Hill or its equivalent at a normal rate without stopping was considered to be healthy, whereas the remainder were separated into those with definite evidence of I.H.D. and those without on the basis of the following:

(1) Clinical history of angina pectoris, coronary insufficiency or coronary thrombosis.

(2) A negative physical examination (cases with anaemia, thyrooxicosis, or syphilitic aotitis were excluded from this study).

(3) An electrocardiogram taken before and immediately after maximum effort, using Wood's (1958) criteria of ischaemic change, namely an ST depression measuring 1 mm or more below the level of the atrial T wave, remaining flat or sloping downward for at least $0 \cdot 08$ sec. On every patient, 6 limb leads and 6 praecordial leads were used. On the whole, the ECG findings correlated well with the clinical history; no unique patterns such as Somers and Rankin (1962) described as present in Africans, or Pyke (1963) in West Indians, were present among St. Helenians.

MILD PRESENTING SYMPTOMS

Some 20 years after first describing the disease, Herrick, during the Harvey oration admitted "into the cadre of acute and subacute cases many milder atypical types in which the supposedly cardinal symptoms may be lacking". It is not generally realized how common silent coronaries are (Lindberg *et al.*, 1960; Master & Geffer, 1964), so that the situation encountered in St. Helena may not be unusual. It was hoped that by selecting a small group of patients with suspected I.H.D. by the procedure mentioned above there would be sufficient time to screen them thoroughly and so decide definitely whether they were healthy or not. Unfortunately this was not always realized. Two men who died of coronary thrombosis (established at post mortem), had been classed as doubtful I.H.D., but after detailed questioning and examination, they were reclassified as normal. The following example illustrates this mild symptomatology.

Mr. R. T., age 75, said that he rested a few times going uphill because he became shortwinded, by which he meant that he felt a burning pain situated below the lower sternum that took his breath away, although adding that he was not forced to rest with the pain. When asked how fast

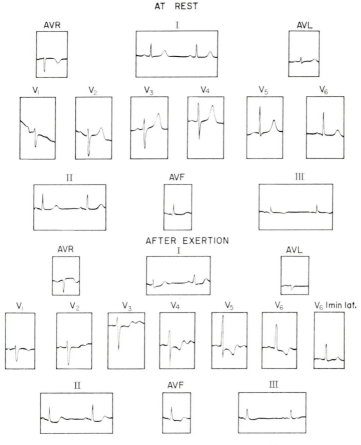

Fig. 3.1. ECG before and after maximum effort.

he went uphill, he said he could run, but not for long, and to demonstrate his ability he ran 15 times up and down a steep 15 yard slope beside the clinic. While taking an ECG immediately afterwards I asked him if he felt any pain and he said nonchalantly, "no pain, but I feel as if the shortwindedness was coming on: I could have run on some more, but I thought to stop before it came on". The tracing (Fig. 3.1) shows obvious ischaemic change.

In an effort to achieve a more objective method of assessing I.H.D., post mortem specimens were sent to the Government Pathological Laboratory in Cape Town where Professor Turner kindly undertook histological examination. While this provided useful confirmatory evidence (stenosed coronary arteries were a common finding), there were insufficient deaths during the period of investigation to derive incidence figures from this source alone.

Whenever myocardial infarct was found at post mortem, the individual was included in the group with I.H.D. provided that he had been examined before death but regardless of the previous diagnosis.

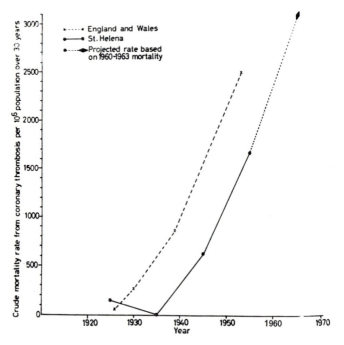

Fig. 3.2. Mortality rate from I.H.D. standardized for every million of the population living over 30 years of age. Comparison between St. Helena and England and Wales.

As an additional guide to the frequency of I.H.D., a crude mortality rate was calculated from the death certificates. All deaths between 1920 and 1963 registered as coronary heart disease, coronary arteriosclerosis, myocardial infarct, coronary thrombosis or angina pectoris were extracted from the death registers, which contained copies of certificates. They are expressed as a crude mortality rate standardized for every million of the population living over the age of 30 (Fig. 3.2). Over this age, and over the period 1920–60, the St. Helenian population had virtually the same age and sex structure as England and Wales, allowing a comparison of the mortality rates in both countries to be made without further adjustment. Although the level of accuracy of previous diagnosis is unknown, all deaths over this period were certified by a medical practitioner.

RESULTS

Of 1500 people examined over the age of 30 (85·7% of the population in that age group) 19 men and 8 women had unequivocal evidence of I.H.D., while for 39 men and 60 women the evidence was uncertain. In addition, 4 men died of myocardial infarct during the period of investigation before they had been examined and were therefore not included in these results. Table 10 shows minimum and maximum prevalence estimates by age, taking the minimum to be that group who were uncontrovertably affected and the maximum to include those people with doubtful evidence of I.H.D. but who could not confidently be considered normal. Assuming people with uncertain evidence of I.H.D. to be abnormal, the prevalence was 7·7% of all males, and 9·1% of all females, but, if all doubtful cases were included with normals, which accords with the recommendation of the criteria committee of the New York Heart Association (1964), the prevalence declines to 2·5% for the males and 1·1% for the females.

ASSOCIATED VARIABLES

The sample was segregated by the presence or absence of I.H.D. and the differences for selected variables tested by X^2 (Table 11). For males, the differences between those with and those without I.H.D. were significant ($P = 0·05$) for the duration of wearing

63

TABLE 10. Individuals Over 30 with Normal Cardiac Function, Definite I.H.D. and Uncertain I.H.D. by Age and Sex Data for 750 Males and 754 Females

Age group (years)	Males					Females				
	Affected	Uncertain	Total	Frequency (%) Min.*	Max.†	Affected	Uncertain	Total	Frequency (%) Min.	Max.
30–	0	0	167	0	0	0	2	208	0	1·0
40–	1	1	190	0·5	1·1	1	6	165	0·6	4·2
50–	8	18	203	3·9	12·8	2	14	161	1·2	9·9
60–	6	12	121	5·0	14·9	4	20	123	3·3	19·5
70–	4	6	53	7·5	18·9	1	16	69	1·5	24·6
80–	0	2	15	0	13·3	0	2	25	0	8·0
90–	0	1	1	0	100·0	0	0	3	0	0
TOTAL	19	40	750	2·5	7·9	8	60	754	1·1	9·0

* The minimum frequency is calculated for individuals with definite I.H.D.
† The maximum frequency is calculated for individuals with uncertain and definite I.H.D. together.

TABLE 11. COMPARISON OF PATIENTS OVER 40 WITH AND WITHOUT DEFINITE ISCHAEMIC HEART DISEASE FOR SELECTED VARIABLES: DATA FOR 583 MALES AND 546 FEMALES

Variables	Males			Females		
	I.H.D. Present	I.H.D. Not present (%)	Significance P level	I.H.D. Present	I.H.D. Not present (%)	Significance P level
Total population	19	564 3·3		8	538 1·5	
Shoe-wearing: < 12 years	6	310 1·9	0·052	1	161 ·6	N.S.
> 12 years	12	248 4·6		7	363 1·9	
Marital status: Single	0	85 —	0·045	1	73 1·3	N.S.
Married	19	471 3·9		7	454 1·5	
Exercise: < 2 miles	10	131 7·1	0·008	5	450 1·1	N.S.
> 2 miles	9	410 2·1		2	67 2·9	
Occupation: Sedentary	3	141 2·1	N.S.	1	33 2·9	N.S.
Other	16	418 3·7		7	500 1·4	
Arcus senilis: Present	9	279 3·1	N.S.	5	203 2·4	N.S.
Absent	8	261 3·0		3	310 1·0	
Smoking: < 5 per day	10	291 3·3	N.S.	7	527 1·3	N.S.
> 5 per day	8	247 3·1		1	7 12·5	
Body weight: < 140 lb	9	244 3·6	N.S.	2	219 ·9	N.S.
> 140 lb	7	249 2·7		4	207 1·9	
Social class: I and II	3	91 3·2	N.S.	3	77 3·9	N.S.
III and IV	16	473 3·3		5	456 1·1	
Diastolic B.P.: < 95	7	366 1·9	0·066	1	230 ·4	N.S.
> 95	12	186 6·1		6	292 2·0	
Systolic B.P.: < 180	10	478 2·0	0·072	3	345 ·9	N.S.
> 180	9	74 9·2		4	177 2·2	

Note: Differences between total for each group and total population relate to cases where information was incomplete.
N.S. = not significant at 5% level.

65

shoes, marital status, and exercise, while there were no significant differences in respect of occupation, arcus senilis, smoking, body weight, social class and blood pressure. None of these variables reached the level of technical significance for women. Since the variables were interdependent, it was of interest to see what happened after allowance was made for the other factors. Multiple regression analysis, following the same procedure described in the

TABLE 12. ESTIMATES OF SOCIOLOGICAL EFFECTS ON I.H.D.

Independent sociological variable	Males (total 956)		Females (total 819)	
	Partial regression coefficient (b)	Significance (t)	Partial regression coefficient (b)	Significance (t)
Physical activity (mpd)	− 0·093	4·27 ***	− 0·017	1·76
Occupation	0·195	3·28 **	0·095	3·54 ***
Diastolic B.P. (mm g)	− 0·017	3·48 ***	− 0·0003	0·20
Age (in years)	− 0·002	0·45	0·001	1·02
Marital status	0·267	1·88	0·008	0·21
No. of children	0·030	1·33	− 0·004	0·76
Cigarettes per day	0·156	1·61	0·060	1·54
Body weight (lb)	0·002	1·05	− 0·0002	0·37
Systolic B.P. (mm g)	0·005	1·79	0·001	0·08
Years shoe-wearing	0·017	5·64 ***	0·043	7·44 ***
Race	—	—	—	—
Arcus senilis	—	—	—	—
District of residence	—	—	—	—
Social class	—	—	—	—
Child deaths	—	—	—	—

Regression M.S.		$0 \cdot 0792 \times 10^3$	$0 \cdot 1836$	
d.f.	955		918	
Residual M.S.		$0 \cdot 2580 \times 10^5$	$0 \cdot 1798$	
d.f.	941		814	

— Not significant in stepwise test.
* $P < 0 \cdot 05$.
** $P < 0 \cdot 01$.
*** $P < 0 \cdot 001$.

System of Screening:
   Ischaemic heart disease is arbitrarily graded into 0 = absent, 1 = doubtful, 2 = definite angina, 3 = angina with coronary.
   Marital status was arbitrarily graded as single =1, married, divorced, widowed = 2, married more than once = 3.
   Occupation was arbitrarily graded according to the amount of exertion involved, 1 being most and 10 least.

previous chapter, indicated that sedentary occupation, exercise and diastolic blood pressure in men and sedentary occupation in women, were significantly associated with I.H.D., but no association was apparent with cigarette smoking, body weight, fertility or arcus senilis (Table 12). Diastolic blood pressure showed an inverse association with I.H.D.; for the remainder the association was in the same direction. Also there was positive association with duration of wearing shoes.

DISCUSSION

This part of the study was initiated because the St. Helenians appeared to be immune from I.H.D., but as soon as they were examined in a systematic way it became obvious that the original impression was wrong. The mistake was due to a combination of unfamiliarity with a strange culture, a sampling error and mild presenting symptoms.

When it became obvious that I.H.D. did occur, the study was continued to see how common the disease was, and perhaps discover association between it and one or more of the factors that are currently held to be responsible in the aetiology of the disease. These factors are cigarette smoking (Beuchley et al., 1958; Dawber et al., 1959; Doll & Hill, 1956; Schwartz et al., 1966), stress (Rosenman & Friedman, 1963; Russek, 1964), lack of exercise (Morris et al., 1953; Morris & Crawford, 1958; Mann et al., 1964), a high intake of fat, especially animal fat (Bronte-Stewart et al., 1955; Keys, 1953, 1956, 1958) and, more recently, a high intake of sugar (Yudkin, 1957, 1964). Most of these factors are such that one would expect a low prevalence of the disease in St. Helena, yet I.H.D. appears on St. Helenian death certificates almost as often as on British death certificates (Fig. 3.2). In so far as one can compare prevalence estimates of the disease in the living between different populations which are inevitably based on different diagnostic criteria, the St. Helenians resemble communities with a high rate, like the British, rather than communities with a low rate like the Bantu (Schrire, 1958, 1964), among whom coronary thrombosis is still so uncommon that small series of cases are reported in the literature (Kloppers, 1961).

By using stringent criteria it was intended to provide a reliable minimum estimate, but it is difficult to judge how many people should have been classified as affected yet were not. The electro-cardiogram was taken within 30 sec of completion of exercise, probably corresponding to the 2 min record, and so likely to miss these abnormalities seen during the period of exertion (Lloyd-Thomas, 1961) or at 6 min (Master & Rosenfeld, 1964), which may overlook as many as 30% of patients tested (Rogers & Hurst, 1964). This type of error must be to some extent compensated by misinterpretation of the changes in ST segment that frequently occur in normal people during ordinary activities (Hinkle *et al.*, 1964), and by counting symptomless people as normal. Hence it is probably not far off to compare the 5% prevalence among St. Helenian males aged 60–69 with 8·4% of males in Birmingham, England, where approximately similar diagnostic criteria were used (Brown *et al.*, 1957), although by 1959, 16·4% of the same group were affected (Record & Whitfield, 1964).

The uncertainty about the number of missed cases should not affect the analysis of sociological differences since they would be diluted by the larger number of genuinely normal people. Analysis of concomitant variables by means of X^2 and multiple regression showed significant effect for physical inactivity, sedentary occupation, shoe-wearing and low diastolic blood pressure. Because only current estimates were obtained, these factors are as likely to be the result of I.H.D. as the cause. I.H.D. would be expected to restrict physical activity, induce people to wear shoes, take a sedentary occupation, and it might also tend to lower the blood pressure. Nonetheless, diastolic blood pressure apart, these factors are the ones generally considered pathogenic.

Many large epidemiological studies (reviewed by Epstein, 1965) have revealed some interesting but so far inconclusive information. The Bantu, the Yemenite Jew, the Japanese, Chinese, Jamaican and Apache Indian are found to have a low incidence, and, on the whole, the New Yorkers, the English, Finns and Swedes have a high incidence which fits in with the general implication of the present data that the pathogen is associated with Western living although giving no indication which particular feature of Western living is responsible. In these studies no single atherogenic factor has been consistently related to the disease. This may be because the epidemiological method does not provide sufficiently accurate results, or because the initial pathogen is unrecognized or perhaps because different factors are pathogenic to some peoples and not to others or in some environments and not in others. The basis for the different ethnic incidence may be genetical; both Heberden (1772) and Sir William Osler (1910) noted that the disease sometimes runs in families, and this observation has been reconfirmed more recently (McKusick, 1964; Epstein, 1964); however, it does not help to explain the rapid secular change, which most investigators believe to be real and which, if real, requires an environmental explanation. Moreover, the similarity of the rates of increase of I.H.D. in St. Helena and England (Fig. 3.2) suggests that there are few, or perhaps only one main determinant, since it is unlikely that several factors have spread at the same rate in England and St. Helena, whereas it seems not unlikely that there should be the same rate of increase in consumption of sugar, aspirin or cereals. There is some evidence of negative correlation between calcium consumption and I.H.D. (Boström & Widström, 1965).

It is interesting that the St. Helenians, although in general avoiding those pursuits that are currently believed to lead to I.H.D., are not protected from the disease. They go about on foot over hilly countryside, so that they are far fitter than the average European. Just travelling to and from work, the men stated, they walked 5 miles a day and the women $3\frac{1}{2}$. High physical activity is believed to protect the Masai in spite of their high milk diet (Mann *et al.*, 1964) and is a common factor found among most communities with a low I.H.D. prevalence (Lowenstein, 1964), but it does not seem to protect the St. Helenians. There is little of the sort of street life met with in Western society; thus the St. Helenians are such an easy-going, delightful and gentle people that if Captain Cook were to return he could repeat the same description that he gave of them in 1775. The consumption of tobacco is low, the men smoke 6 ± 4 cigarettes a day and almost all the women are non-smokers. The intake of fat as 29% of the calories was lower than in the U.S.A., Canada, and Australia, with Britain's intake at about 38% (Jolliffe & Archer, 1959)—and one-tenth of it was unsaturated marine fat. In so far as this type of discussion is valid, the presence of I.H.D. in St. Helena in the absence of those factors that are held to have high protective value leaves Yudkin's hypothesis that sugar consumption is the primary cause of the disease as a possible explanation (Shine & Barr, 1966). The high level of sugar consumption in St. Helena (over 100 lb per adult per year) fits nicely into this scheme, and compares with 120 lb in Britain and 60 lb among the rural

Bantu (Cleave & Campbell, 1966). These authors have compiled an interesting collection of epidemiological and clinical support for this hypothesis using data of their own and other workers (Cohen, 1963; Yudkin, 1964; Yudkin & Roddy, 1964). However, there are a few conflicting reports (Papp *et al.*, 1965; Little *et al.*, 1965) which may be due to differences in methodology (Yudkin, 1965).

In this study, being unable to relate individual sugar consumption to cardiac function, since dietary data were collected anonymously, a great deal of information was lost that might well have provided a definite answer; as it is, the only permissible conclusion is that the St. Helenians appear to be changing from an underprivileged agrarian society into a more prosperous one, and this is concomitant with a change from a low to a high incidence of I.H.D.

CHAPTER 4

CONGENITAL AND INHERITED DISORDERS

The medical officer, Dr. Wignall, always drunk and nearly killed the Governor by giving unsuitable medicines, his excuse being he had nothing else to give. Dr. Wignall for drunken disorderly conduct placed in the Stocks for one hour and he sung and swore the whole time.

(4 January 1725, *St. Helena Consultations*)

HAVING found a few rare abnormalities soon after my arrival, I decided to examine the entire island to discover how few and how rare they were, probably hoping that it might be possible to interpret any departures from expectation.

METHOD OF EXAMINATION

When people attended the medical clinics (which they did at various times of the day), they were given a routine physical examination limited to an inspection of the gait, posture, face, ears, eyes, mouth, hands and feet; also the character of the voice was noted, the intelligence assessed and the heart auscultated. The fundus was examined with an electrical ophthalmoscope without midriatics whenever patients had poor vision or ocular anomaly, if a sib was known to have retinitis pigmentosa and on all deaf mutes, except one who was accidentally overlooked. All abnormalities were described, photographed and subjected to whatever method of investigation was appropriate and feasible. This system of examination could be carried out rapidly and reliably, and it had the further advantage that it required little undressing or inconvenience to the patient. In most published pedigrees the unaffected individuals are those not known by their relatives to be abnormal or those who failed to come to the physician's attention, whereas the system of examining everyone ensures that the presence or absence of anomaly is indicated with equal reliability.

71

FAMILY HISTORY

Great tact was required to elicit a family history of any inherited disease except Christmas disease, which was the only one that did not carry social stigma, and for this reason the only routine direct questions asked referred to this condition. It was easy to construct pedigrees showing the relationship between individuals, but hearsay evidence of deformity proved unreliable. This caused no error regarding living people, as they would be seen anyway, but it meant that information regarding previous generations and dead children was unreliable. However, as it was possible to check the information from several sources, it was possible to improve the quality of information considerably.

DEFINITION OF CONGENITAL AND INHERITED DISORDERS

The estimation of the frequency of these diseases is so subjective that comparisons between studies are rather like comparisons between the croquet players in *Alice in Wonderland*; for the flamingo hedgehog and deck of cards are about as variable as the physician, the disorder and the population. The following list includes most of the common sources of inaccuracy in deriving frequency estimates.

A. *Variation among physicians*

(i) *Level of diagnosis*. As physicians have different experience, some will record that which others will overlook. Dr. Wignall is an extreme example of variation to be found among physicians, but even today among equally competent doctors, inconsistencies are not uncommon. This was so in Neel's study of Hiroshima and Nagasaki (1958), in which clinical examination of infants was undertaken by physicians specially trained for the purpose, yet re-examination at 9 months revealed examples of achrondroplasia, anal atresia, cleft palate, congenital dislocation of the hip, club foot and many other disorders that had been missed at the original examination (Neel, 1958). This is also illustrated by a family with a congenital oesophageal lesion and tylosis (Shine & Allison, 1966) that had been previously reported at an international gastroenterology congress on account of the familial oesophageal anomaly. However, the tylosis had not been noticed, even by the propositus, himself a physician, until, at the age of 65, the development of oesophageal carcinoma drew attention to it.

(ii) *Diagnostic criteria.* Different physicians, noting the same physical signs, will differ in their readiness to consider the character abnormal or the abnormality congenital. This is inevitable since there are no agreed criteria of normality nor of congenitalness. For example, if a lesion is ascribed to trauma should it be rejected? Should a disease with an obvious infective basis such as leprosy be rejected? In the family with unilateral knock-knees, to be described shortly, at first it did not occur to me to ask if affected individuals were related because the lesion was unilateral, ascribed to trauma and knock-knees were not known to be inherited. It is particularly difficult for physicians to maintain uniform criteria for defects that are arbitrarily separated from normality. How minutely should people be examined? Some traits, like amino-isobutyric-aciduria give no outward sign at all and can only be detected by laboratory investigations; hence, the adage that a normal patient is one who has been insufficiently investigated.

(iii) *System of classification.* Since the sixteenth century when Ambroise Paré first attempted to classify congenital disorders, several systems of classification have been proposed, but none as yet has found general acceptance. Even such apparently unmistakable entities as anencephaly are frequently recorded as "monster" or multiple abnormalities (Morton *et al.*, 1967).

B. *Variation in the presentation of diseases*

(i) *Diseases show varying expressivity in space.* As mentioned above, when a defect is small the point of separation from normality is arbitrary. For example, it is hard to decide when small eyes become microphthalmic, or whether a small defect in the upper lip should be considered as a harelip, or when bow legs are bowed. Schull and Neel (1965) suggest ignoring anomalies less than three standard deviations from the mean, which is an excellent standard except when means and standard deviations are not known, which is the case for many medical parameters at different ages in different races. When searching for such standards (for any race) with which to compare the St. Helenian dwarfs, I was unable to find any that were suitable; no one, for example, has figures relating metacarpal length with height at all ages; nor even means and standard deviations for the lengths of all the long bones, not to mention the width; although there were several good sets of measurements for parts of the body, such as Venning's (1956) for the feet. Such statistics would be boring to compile but of great value; but obviously it would be very difficult to establish means for certain disorders. This is plain when one asks what is the mean number of naevi per person, or what is the mean size of naevus?

(ii) *Diseases show varying expressivity in time.* Most diseases show some fluctuation in expression, for example tylosis regresses seasonally, familial Mediterranean fever is by definition periodic, haemophiliacs have non-haemorrhagic intervals, myasthenia gravis is harder to detect in the morning; and the seasonal variation in incidence of birth defects (McKeown & Record, 1951) provides a more general example.

(iii) *Diseases vary in time of onset.* Anencephaly is apparent at birth, pyloric stenosis at 2 weeks, phenylketonuria at about 1 month, Tay Sachs disease at 5 months, Duchenne muscular dystrophy in childhood, and so on. Clearly, examination of a population at birth would not be expected to detect Tay Sachs disease.

(iv) *Variation in latency.* (a) Traits may be physically concealed; a moustache can hide a harelip or a dark skin purpura, and medical treatment may remove the deformity or cure the disease. (b) Traits may be biochemically concealed and require environmental or medical provocation before they can be detected. Popphyria cutanea tarda requires sunlight, pseudo-cholinesterase deficiency requires suxamethonium and galactosaemia requires milk.

(v) *Variation in survival of affected individual.* Since the probability that an abnormal child will survive varies from country to country, prevalence differences may merely reflect the standard of medical care. Thus a doubling of the reported population frequency of mongolism in Britain over the past two decades has nothing to do with the birth incidence, which has not altered. Part of the reason why there were only two mental defectives among 6687 Brazilians (Mi *et al.*, 1965) compared to 43 cases in St. Helena, is that Brazil has a high mortality and St. Helena a low mortality rate. Furthermore, many societies have practised infanticide for abnormality (e.g. Germany) or for twin births (e.g. Japan), while others have taken the opposite view and considered abnormality a mark of distinction. This is generally believed of albinos among American Indians, although apparently not so among the Hopi (Woolfe &

Grant, 1962). If albino males are given mating privileges, clearly the albino gene could spread rapidly; another example is provided by the Hyabites, among whom polydactyly was so general that any child born with five fingers was regarded as deviant and sacrificed (Boinet, 1898).

C. *Variation in total population*

In this text, frequency is defined as the number of cases of a disorder occurring in a defined population in a given time. However, it is not at all clear how to define a population. Should the boundary correspond to the arbitrary census divisions or to biological divisions and if so, which ones? Are the boundaries fixed at the beginning of the study, or are they extendable if an interesting group of patients is found just beyond the country border? In an isolate the problem is somewhat easier but still there is some ambiguity. Should one include only those St. Helenians living in St. Helena, ignoring the larger part that live elsewhere? And this presupposes that it is possible to define a St. Helenian. Moreover, in any sample of a larger population there is a serious risk that it may be biased. An obvious source of difficulty in comparing prevalence estimates is due to the stratification of societies. The Population Genetics Research Unit conducted a survey into sex-linked traits around Oxfordshire and found that many families with haemophilia had moved to Oxford to be near Drs. Macfarlane and Biggs (Kerr, 1965). Similarly, the high sex ratio among people with congenital disorders on Ascension Island is attributed to the high sex ratio among the total population which is the result of the tendency of employers to employ men who were rarely allowed to bring their families with them.

Even if all the foregoing difficulties could be overcome it is doubted whether any statistic accurately tests population frequency differences. The X^2 test assumes independence of the original observations, which, among related individuals, is an invalid assumption (Cotterman, 1952). To minimize errors arising from these sources the following procedure has been adopted.

(1) Every St. Helenian was examined and questioned personally under similar conditions thereby ensuring a reasonably consistent level of observation, diagnosis and classification.

(2) St. Helenians were defined as those people who were born in

75

St. Helena if they considered themselves St. Helenians and were currently living either on St. Helena or Ascension Island.

(3) An abnormality was included if it was:
 (a) obvious on clinical examination,
 (b) extensive enough to cause symptoms,
 (c) genetically determined.

The criteria for inclusion are hard to define without any ambiguity, and some may disagree with the present choice. To permit an alternative classification, disorders are described in detail whenever they are borderline or unusual. By this means the level of diagnostic stringency and the basis for the present classification will be obvious (Table 13).

TABLE 13. THE KINDS OF CONGENITAL AND INHERITED DISORDERS OBSERVED AMONG 4259 ST. HELENIANS BETWEEN 1960 AND 1962

1. SINGLE MALFORMATIONS

	Total
A. *Musculoskeletal system*	
Accessory fingers and toes	4
Arthrogryposis and club foot	1
Brachydactyly (a) "pseudobrachydactyly"	6
(b) Julia Bell type A and D	7
(c) Julia Bell type A3	1
(d) Julia Bell type D	16
(e) Julia Bell type E	1
(f) Variant of D	1
Club foot all types	7
Congenital dislocation of the hip	2
Dwarfs with infantilism	3
Elbow congenitally stiff	1
Multilocular cyst of femur	1
Muscular dystrophy	3
Osteochondrodystrophy	1
Synostosis 1st and 2nd toes	1
Unilateral familial knock-knees	7
B. *Respiratory system*	0
C. *Cardiovascular system*	
Congenital heart disease—type undetermined	3
D. *Haemic and lymphatic system*	
Christmas disease	21
Congenital hyperbilirubinaemia	4
Purpura (thrombocytopenia) (at least)	46
E. *Digestive system*	
Harelip and cleft palate complex	12
Hirschsprungs disease (diagnosed in Cape Town)	1
Malabsorption syndrome (coeliac disease)	1
Pyloric stenosis	1
Thyroglossal cyst	1

TABLE 13 (*continued*).

F. *Urogenital system*   *Total*

Hypospadias	1
Unilateral gynaecomastia (large enough to require surgical removal)	1

G. *Nervous system*

Friedreich's ataxia	2
Microcephaly	2
Epileptics on treatment	11
Mental deficiency alone	34
Mental deficiency with rasping voice	7
Mongolism	5
Multiple neurofibromatosis	1
Nystagmus alone	6
Nystagmus with voice defect	2
Nystagmus secondary to ocular abnormality	9
Palilalia echolalia syndrome	2
Spina bifida with meningocoele	1
Spasticity alone	6

H. *Organs of special sense—Eye*

Microphthalmos	3
Choroiditis	3
Coloboma iridis	1
Partial coloboma of optic nerve with ptosis, and torticollis	1
Corectopy	11
Ptosis	13
Optic atrophy	4
Strabismus	61
Pterygium	92
Retinitis pigmentosa alone	1
Retinitis pigmentosa with deaf mutism *	1 (or more)

I. *Organs of special sense—Ear*

Accessory auricle	23
Deaf mutism alone	4
Deaf mutism with retinitis pigmentosa *	1 (or more)
Deaf mutism with cataract	3

J. *Integumentary system*

Albinism	5
Keratosis pilaris	3
Ichthyosiform erythrodermia	3
Vitiligo	3
Haemangioma	8
Dermatosis papulosa nigra	32
Familial St. Helenian fever	15

2. MULTIPLE MALFORMATIONS

Anencephaly with cervical meningocoele, agenesis right lung, agenesis right thumb, imperforate anus, indeterminate sex	1

* Double entry.

	Total
TABLE 13 (*continued*).	
Atresia ani with rectovaginal fistula and congenital heart disease and microcephaly	1
Talipes with pale fundi, mental deficiency, and spasticity	4

The death register includes the following congenital malformations:

Anencephaly	7 cases between 1860 and 1880, none since until 1961. Three recorded since.
Hydrocephaly	1860–80: 6 cases; 1 since 1940.

BRIEF CLINICAL DESCRIPTION OF DISORDERS WITH FEATURES OF PARTICULAR INTEREST

Appended pedigrees are drawn according to the Galton Convention. An interrupted line indicates uncertainty that the pedigree is as shown.

Albinism

Five generalized albinos were seen and a further sixteen recently deceased were ascertained (Fig. 4.1). Fitzpatrick and Quevedo (1966) doubt that adult albinos commonly show a pink-light reflex, "a common misconception about the colour of the iris and the pupillary reflex in persons with oculocutaneous albinism may be based on inferences from the pink eyes of albino mice and rats". However, it was clearly seen in the two albino children as well as

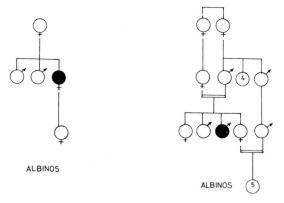

ALBINOS

ALBINOS (5)

Fig. 4.1

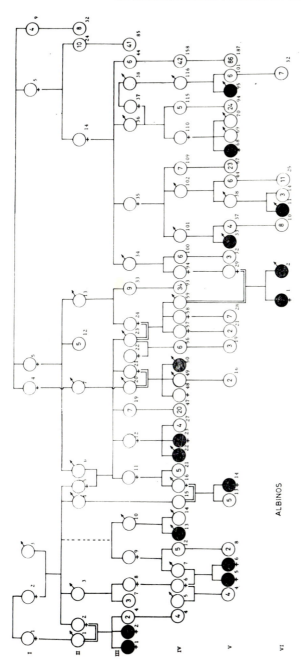

Fig. 4.1 Pedigree of albinos.

79

the three adults (Fig. 4.2), and it was sufficiently obvious to the St. Helenians to earn one adult albino the nickname of "Ebby Boom Bang" (Chapter 1). An accentuated light-reflex is a feature of the Chediak–Higashi syndrome (Chediak, 1952; Higashi, 1954), although this diagnosis is unlikely to apply in this instance since few patients with this syndrome have lived longer than 7 years. The St. Helenian albinos all showed the typical features of fundal and iris depigmentation with nystagmus and photophobia—they had yellow to light brown hair, and skin that was paler than their sibs but slightly tanned and reddened over the exposed areas. All had reduced intelligence.

A history of haemorrhagic tendency was obtained from III$_3$ (see Fig. 4.5) but the bleeding time, clotting time and tourniquet tests were normal, which is similar to the findings of Hermansky and Pudlak (1959) and Verloop et al. (1964).

Three individuals showing some depigmentation were not counted as albinos since they did not fulfil all Froggatt's (1960) criteria, namely retinal depigmentation, congenital nystagmus, translucent irises and depigmented hair.

It is clear from the pedigree plates that the trait behaves as an autosomal recessive. There is an increased frequency of cousin marriage among parents of affected individuals, the segregation ratio is that expected of a recessive mendelian trait and the sex ratio is approximately unity.

Hogben's (1931) a priori test of a recessive hypothesis on pooled sibships containing at least one observed albino (Table 13a).

TABLE 13A

Number in sibship (s)	Number of sibships (n_s)	Affected		Variance of expected number ($n_s o_s \sigma^2{}_s$)
		Observed	Expected	
2	1	2	1·1428	0·12245
3	2	2	2·5946	0·32594
4	1	1	1·4628	0·42005
TOTAL		5	5·2002	0·86844

The difference between observed and expected = 0·2 is less than one S.D. = 0·9 of the expected number.

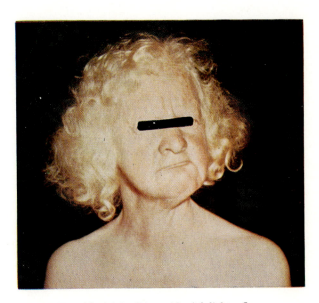

FIG. 4.2. Adult albino with pink light reflex.

Accessory digits (polydactyly)

Three people were seen with an extra digit, and another man had a scar that he alleged was the result of removal of an extra finger. Since the St. Helenians tend to remove them, the true frequency is probably higher than found. This trait apparently shows some racial variation (Stevenson *et al.*, 1966; Handforth, 1950).

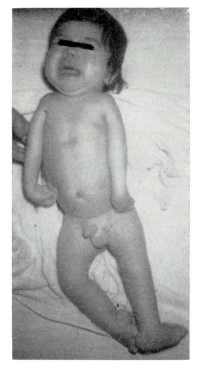

FIG. 4.3. Arthrogryposis multiplex congenita.

Arthrogryposis multiplex congenita

The youngest child of an otherwise healthy sibship of 5 was born with joint stiffness, rotation deformities and muscle-wasting characteristic of this disease. He was never able to move his limbs, he showed little evidence of intelligence, failed to thrive, and died in status epilepticus at 5 months. There were no other affected relatives, but his father's mother's brother had osteochondrodystrophy.

This disease had not generally been considered genetical

(McKusick, 1961). Mead *et al.* (1958) obtained negative family histories from 40 cases; and 4 pairs of monozygotic twins; 2 in his series, and 2 reported by Hillman and Johnson (1952) were discordant for this disease. However, Frischknecht *et al.* (1960) reported a sibship in which 3 members were affected, and Weissman *et al.* (1963) reported an Arab family apparently showing a recessive inheritance.

Brachydactyly

Defining brachydactyly as a marked diminution in the length of the hands or feet or of one or more of their component rays, 32 people were seen with brachydactyly, giving a population frequency of a little over 1:130. They fell into 6 distinct groups, 3 of which corresponded to Bell's (1951) type A_3, D, and E; in addition, one family showed a combination of type A_3 and D in which one or both anomalies were manifest in the heterozygote (Fig. 4.4), in another all the terminal phalanges were small with short, broad nails in all digits; and, finally, there was a unique syndrome of pseudobrachydactyly associated with dwarfism, which was the only variety to show recessive inheritance.

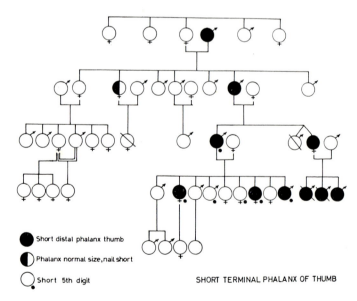

● Short distal phalanx thumb

◑ Phalanx normal size, nail short

○ Short 5th digit

SHORT TERMINAL PHALANX OF THUMB

FIG. 4.4. Pedigree of brachydactyly type A_3 and D.

Christmas disease

A large family was found with Christmas disease, which was first reported by Wilkinson (1938) as probable haemophilia. It is difficult to be certain of the number affected because the symptoms were mild, and simple laboratory tests gave normal results. The bleeding time, clotting time, tourniquet test and blood film examinations were usually normal. Dr. T. Geddis, the medical officer on Ascension Island, treated IV_{128} (Fig. 4.5) for an epistaxis in which blood oozed continuously and uncontrollably from the outer lateral surfaces of the nose for 3 days. On the second day Dr. Geddis collected blood from the patient and from himself as a control, and allowed them to clot in a clean ordinary test-tube which was held in the hand as an incubator. The patient's blood took 33 min to clot and his own took less than 5 min to clot. The first opportunity for more orthodox laboratory investigation occurred in 1965, when blood was taken from V_{117} on Ascension Island and flown to Oxford where Drs Biggs and Macfarlane kindly performed factor IX assay and other tests. They found 6% of factor IX, which compared to a normal St. Helenian control taken at the same time, was considered consistent with the diagnosis of Christmas disease. It was arranged that blood should be sent from V_{118} as well, but he developed a severe gastro-intestinal haemorrhage and was sent to Cape Town for investigation, where, according to a personal communication from Dr. Macfarlane, a low level of factor IX was found by Dr. H. Nossel.

Apart from this episode of gastrointestinal haemorrhage the islanders had been remarkably free of internal or joint haemorrhage, their usual symptoms being "nose bleeds" and excessive haemorrhage after minor injury and dental extraction, which seemed to become milder with advancing age. It is not known whether nose bleeds refer to the extraordinary type that Dr. Geddis observed. Only two unequivocal examples of excessive haemorrhage were personally observed, both with haemorrhage lasting several days following the extraction of a single tooth. The haemorrhage commenced about an hour after extraction, and after establishing control, repeatedly began afresh after an interval of 1 hour to 1 day.

Four members of this family have died from haemorrhage; III_9, when 20 years old, died of haemorrhage from the urethra after catheterization; his brother III_{10} died at the same age after cutting his foot; IV_3, aged 8 years, died after cutting his foot on a bully-beef tin; V_2, aged 18 years, died of haemorrhage after a motor-cycle accident, and V_{76}, a 13-year-old girl, died of haemorrhage after tonsillectomy.

The pedigree in this family looks somewhat odd, but it has been recorded as it was obtained from the clinical histories. The probable reason for the irregularities is that the histories are unreliable since most relatives of genuine bleeders seemed eager to have the disease. For example it was not certain whether V_{76}'s father was affected, which is essential if his daughter is to be considered homozygous. The problem was accentuated due to the tendency of the disease to become milder with age, so that V_{10-15} give a convincing history of bleeding up to the age of 30 (corroborated by Wilkinson's report), but they claim to have experienced no abnormal bleeding over the past 25 years. However, there is no doubt about the diagnosis of Christmas disease since low factor IX levels have been demonstrated in two members. In view of the mildness and inconstancy of the symptoms, the apparent male to male transmission and the presence of affected females, it would have been easy to dismiss the entire story as fiction had the laboratory analysis been performed by anyone less reliable.

Mildness and inconstancy of symptoms among patients with the same level of factor IX have been well established (Quick, 1957; Moor-Jankowski et al., 1957; McKusick & Rappaport, 1962).

The presence of affected males with affected sons does not rule out Christmas disease; it merely rules out sex-linked transmission, for there is no reason why factor IX deficiency should not result from an autosomal mutation, as has recently been described for haemophilia (Hensen et al., 1965): although in an inbred community it is quite likely that the wife of an affected male is the carrier.

The presence of affected women has three possible explanations. Firstly, normal women frequently give false histories of bleeding tendency (Macfarlane, 1962; Owen et al., 1964). Secondly, they could be examples of gene expression in the heterozygote, which is not uncommon (Hardisty, 1957; Simpson & Biggs, 1962), or they could be examples of homozygosity, but this has never been previously reported. It would have been easier to accept this possibility with some confidence had the exact relationship between V_{76}'s parents been known (they were merely said to be related) and had her father showed definite evidence of bleeding instead of giving a history of it referring to his youth. Nonetheless, Wilkinson considered the father to be affected and reported that the females in this sibship had been seen to bleed abnormally; moreover, V_{76} did die of haemorrhage after tonsillectomy. Tonsillectomy is a dangerous and often fatal way of detecting mild clotting defects, although most of these deaths occur in people with normal blood coagulation (Krahl, 1955; Tate, 1963).

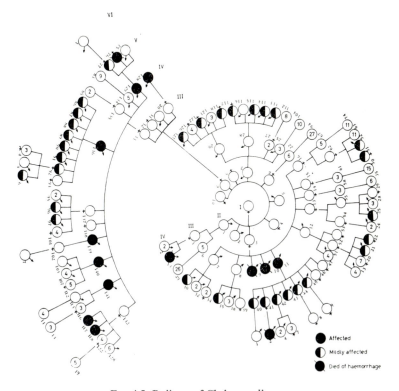

FIG. 4.5. Pedigree of Christmas disease.

Club foot

Only talipes equino varus is included, since the diagnosis tends to be unreliable if all foot deformities are recorded. The diagnosis is also unreliable in the newborn who were therefore excluded. Two neonates with talipes calcaneo valgus had normal feet when re-examined at 3 months. The observed frequency of 1 : 663 is a little more than the usually quoted figure of around 1 : 807 live births (Wynne-Davies, 1965).

85

Congenital dislocation of the hip

Two individuals seen with this diagnosis were both female, one had bilateral dislocation, while in the other it was unilateral. They were not closely related and had no affected relatives. One borderline case was not counted because she was now aged 12 and entirely normal. She had been considered affected on the basis of clinical examination and X-ray when newborn, and had been placed in a hip spica for 3 months. Since diagnosis in the newborn can be difficult, and as there was now no direct evidence available, it was considered safer to reject this case.

Colour blindness

All literate school children (831) were tested with Ishihara test-plates (15th complete edition, 1960) by Mrs. May Young, the assistant health visitor. $2 \cdot 3\%$ of the male children were colour blind of the deutan type. If the assumed composition of the population is correct (Chapter 1), a higher incidence of colour blindness might be expected since Africans show a frequency about $2 \cdot 5\%$, the Chinese about 6% and the English about 8% (Kalmus, 1965).

Deaf mutism with cataract

Three sibs, aged 75, 66 and 63 years, were completely deaf; two were also mute and all had bilateral cataracts obscuring the fundus, although none of them admitted defective vision.

It is possible that this pedigree represents a chance association of anomalies, or a unique mutation, but as they are distantly related to the family with retinitis pigmentosa with deaf mutism and as the majority of deaf mutes with retinitis pigmentosa over 40 years of age develop cataract (Hallgren, 1959), the whole family may have the same disease.

Deaf mutism with retinitis pigmentosa

The only certain example of this disease was an intelligent lady, aged 35 (Fig. 4.6, IV_2), who also had right bundle branch block and the classical fundal appearance of retinitis pigmentosa. Three other members of the family were deaf mutes and may have had retinitis pigmentosa, but the fundus of V_{112} was not examined, the fundus of V_{18} was obscured by corneal opacity attributed to gonococcal conjunctivitis, and VI_{36} had a normal fundus but as he was only 8 years old, he may have been too young to show the typical changes (Hallgren, 1959). This patient also had a bony mass in the hypogastrium, 13 cm in diameter, which was considered to be a lithopaedion (Fig. 4.7), although a teratoma is an alternative diagnosis.

It would seem quite likely that all the deaf mute members of this family also had retinitis pigmentosa.

One old lady was seen with retinitis pigmentosa without any

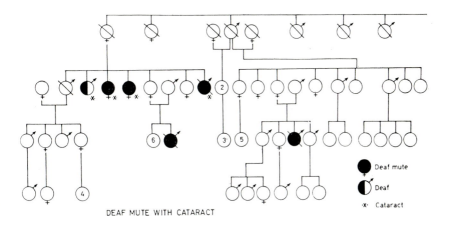

FIG. 4.6. Pedigree of deaf mutes with cataract.

other anomaly. I was unable to obtain any history of affected relatives or examine any of them, but she was definitely not related to the family mentioned above. I have recently heard that she is related to one of the women who went to Tristan da Cunha in 1827, which is of interest since Sorsby (1963) has reported recessive retinitis pigmentosa in 4 out of the 259 present inhabitants.

Dermatosis papulosa nigra

This common anomaly was manifest in the heterozygote, women were apparently more commonly affected than men, and it was not observed in any children. This trait is occasionally found among Africans, but rarely among Europeans (Sutton, 1956).

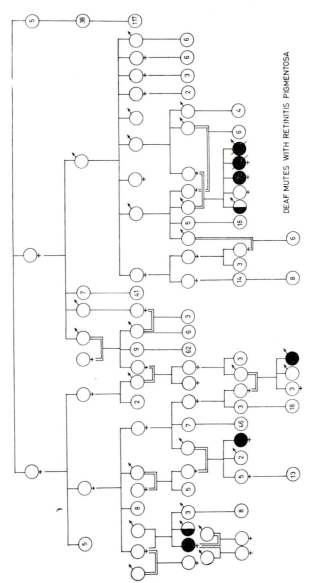

DEAF MUTES WITH RETINITIS PIGMENTOSA

FIG. 4.7. Pedigree of deaf mutism with retinitis pigmentosa.

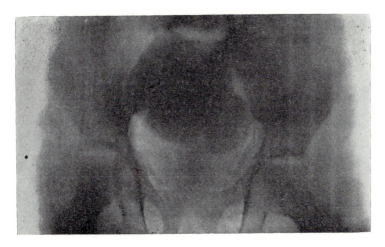

Fig. 4.8. Straight A.P. abdominal X-ray in boy with lithopedion.

Dwarfs with pseudobrachydactyly

Six dwarfed individuals who appear to be examples of a hitherto undescribed syndrome were examined. They were normally proportioned, about 4 ft 4 in. tall, but their hands and feet appeared unusually small even relative to their small stature. However, measurements of the length of the small bones of the hands and feet compared to the remainder of the body showed that no bones were missing or abnormal in structure, but the entire skeleton was proportionately reduced in size (Table 14). The illusion of brachydactyly was created because the small bones of the hands and feet were not reduced in width to the same extent that they were reduced in length. No abnormality was detected in the cardiovascular, respiratory, alimentary or central nervous system, and their intelligence and mental state were normal. All showed normal adult secondary sexual characteristics, and one man and two women had produced children. The hair and skin were of normal colour and texture, there were no striae, and the face was not unduly fat or wrinkled. Except for two individuals who had short, broad nails, physical examination, electrocardiogram and urine analysis were all normal.

No similar case reports were found in the literature. Hanhart's (1925) dwarfs were shorter, fatter and had no secondary sexual development; they probably had pituitary deficiency. There was no evidence of pituitary deficiency in this family and pure growth

TABLE 14. BODY MEASUREMENTS IN CENTIMETRES OF THE DWARFS WITH PSEUDOBRACHYDACTYLY WITH A NORMAL ST. HELENIAN AND NAVAL RECRUITS (ROBERTS, 1957) AS CONTROLS (MEAN VALUES)

	Normal St. H.	Normal U.K.	F.B.	E.B.	C.F.	Ei.T.	I.F.	E.T.
Sex	F	M	F	M	M	F	F	F
Age	55		24	19	40	39	18	34
Height	168	173	132	133	137	134	124	135
Crown-pubis	73	85·5	65	67	72	70		71
Pubis-heel	85	88	67	69	66	64		64
Arm span	158	181	129	137	139	134	134	134
Head circumference	51	55	51	53·5	55	55	53	53
Hand length *	18		12·5	13	16	14	12·5	14·75
Hand span	20		14	17	19	15·25		17·5
Index finger †	7·75		5·75	4·5		5·5		
Upper arm ‡	32	33	27·5	30	28	28	25	27
Lower arm §	25	26	19	21	22	19	18	19·5
Clavicle	15		13·5	13	16	13	16·5	12·5
Sternum ‖	14		16	16	17	16	14·5	15
Inter-spinous distance (A.s.i.s.)	22	25	22·5	20	24	22·5		22
Gtr. troc.—Add. tubercle	35		30	32	25	29		30
Tib. tub-med. malleolus	33		27·75	28	28	25	24	26·5
A.s.i.s.—tib. tubercle	52		44	46	40	46		45
Foot breadth	10	10	10·5	11	12	10	10	10·5
Foot length	25·5	26·5	17	21·5	23	20	18	20
Big toe length (from interphalengeal joint)	7·5		5	5	6	5·5	5	5·5
Thumb nail length			0·6	·8				
Ter. phalanx (from interphalengeal joint)			1·6	2·7				
Breadth head 2nd M.T.C.—X-ray	1·5	1·5	1·4	1·5	1·6	1·5	1·5	1·5
Breadth midshaft 2nd M.T.C.—X-ray	1·1		·5	·8	·7	·7	·6	·7

* From distal wrist crease. † From palmar crease. ‡ From acromion to lat. epicondyle of humerus.
§ Lat. epicondyle humerus to radial tubercle. ‖ Intrasternal notch to ziphisternal junction.

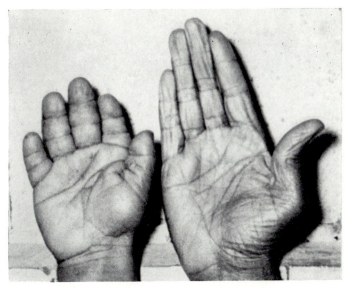

FIG. 4.9 (a). Dwarfs with pseudobrachydactyly.

hormone deficiency is incompatible with normal bone age found in I.F. aged 18 and E.B. aged 19. The Amish dwarfs had metaphyseal dysostosis and sparse hair (McKusick, 1964). Farabee (1903) and Drinkwater (1908) described dwarfs with short legs and brachydactyly due to rudimentary or absent middle phalanges. Primordial dwarfs have characteristic joint stiffness and walking difficulty which these people did not show, and achondroplasia is characterized by a striking facial appearance and body disproportion. Brachydactyly may be observed in Turner's syndrome, pseudohypoparathyroidism and Morquio's disease, but Turner's syndrome is obviously untenable; a normal intelligence, absence of metabolic defect or ectopic calcification makes the diagnosis of either variety of pseudohypoparathyroidism unlikely, and there was none of the bony distortion found in Morquio's disease; and, in addition, in the present family, the metacarpals retain their normal size relative to each other.

It is clear from the pedigree plate that the trait is manifest in the homozygote. Parental consanguinity is more frequent than expected, and the segregation ratio does not differ significantly from the expected 3:1 mendelian ratio (Table 15).

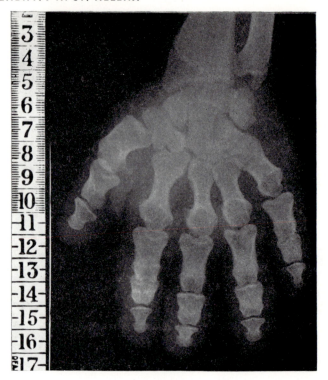

FIG. 4.9 (b).

TABLE 15. A PRIORI METHOD OF TESTING A RECESSIVE HYPOTHESIS ON SIBSHIPS
CONTAINING CASES OF PSEUDOBRACHYDACTYLOUS DWARFS (Hogben, 1931)

Number in sibship (s)	Number of sibships (n_s)	Dwarfs		Variance of expected number ($n_s o_s \sigma_s^2$)
		Observed	Expected	
3	1	2	1·2973	0·26297
6	1	1	1·8248	0·77595
7	1	2	2·0196	0·97024
9	1	1	2·4328	1·38020
11	1	1	2·8710	1·80530
TOTAL	5	7	10·4455	5·19466

The difference between observed and expected = 3·4455 is less than two
standard deviations = 4·6 of the expected number.

92

Fig. 4.9 (c)

Dwarfs with infantilism

A family of dwarfed individuals who showed reduced bone age and absence of secondary sexual characteristics was seen. They were normally proportioned. III_{30} (Fig. 4.12) was aged $17\frac{1}{2}$ when seen, although she looked more like a normal girl aged 9 years. When she was 13 years old there was slight breast development, but the breasts have not altered in size since then, and she has never menstruated. I have recently heard that she has still not menstruated and is now 20 years old.

On examination no abnormalities were found apart from her small stature and arrested sexual development. It is possible that she may have a low intelligence; but as she had not been to school this was difficult to assess: she had a fair memory and was able to read and write. She certainly had a childish personality; when living in a community home she showed shallow and labile emotions and spent most of her time teasing the older inhabitants.

93

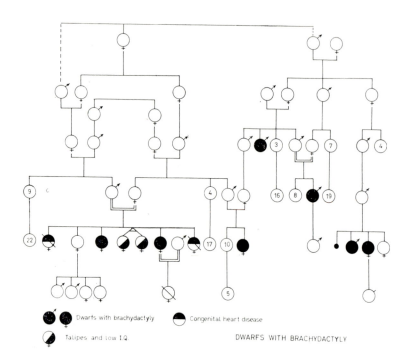

Fig. 4.10. Pedigree of dwarfs with pseudobrachydactyly.

Examination of the central nervous system revealed no other abnormality.

Sexual development. The breasts were very small and immature, with small areola. The nipples were 17 cm apart, and the chest circumference at that point was 66 cm. There was no axillary hair, and only very fine scant pubic hair, the clitoris was normal and the hymen imperforate. The uterus, felt per rectum, appeared normal in size, but no ovaries were felt.

There was no neck webbing, no increase in carrying angle, the hair line was normal and there were no other obvious abnormalities. The body measurements indicated normal proportions and are given in detail in Table 16.

Investigations. X-ray: X-rays of hand and wrist showed a 4–5 year delay in bone age.

TABLE 16. BODY MEASUREMENTS (IN CENTIMETRES) ON INFANTILE DWARFS

Pedigree No.	III_{29}	III_{30}	III_{33}	III_{32}
Sex	F	F	M	M
Age	17·5	14·5	13·5	9
Body weight	65 lb			
Head circumference	52	51	52	52
Height	131	131	147	116
Crown pubis	64	67	70	60
Pubis heel	67	66	78	56
Arm span	138	130	141	120
Hand span	16	17		
Hand length	16·5	15		
Middle finger length	7	6		
Lower arm	21	19·5		
Upper arm	29	27		
Clavicle	11	12·5		
Sternum	11	12		
Inter spinous	19	17		
Gtr. troc.—Add. tub.	30			
Tib. tub.—med. mal.	29	27·5		
A.s.i.s.—heel	76			
A.s.i.s.—tib. tub.	44	42		
A.s.i.s.—knee joint	41			
Foot length	20	20		
Foot breadth	9	10		
Big toe	6	5		

Measurements are taken from the same points as indicated previously.

Urine analysis: The test was kindly performed by Professor C. E. Dent who reported:

> Total nitrogen 455 mg %
> Cystine—normal
> No reducing agents
> Protein—none detected
> Low calcium (Sulkowitch test)
> No ketones (Rothera test)
> No phenylpyruvic acid (Ferric chloride)
> No keto acids
> No porphobilinogen
> Normal amino acid chromatogram

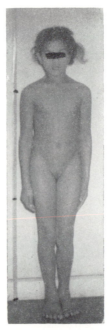

Fig. 4.11. Infantile dwarf (III_{30}).

An aliquot of 24-hr urine was taken to Dr. E. Ross who kindly performed steroid assay. He reported $2 \cdot 5$ mg 17 ketosteroids per 24 hr and $8 \cdot 3$ mg 17 ketogenic steroids per 24 hr.

There was no opportunity to examine the remainder of the family in the same detail, although they clearly segregated into normal and affected.

III_{23} was normal in height, with normal secondary sexual development, and normal intelligence.

III_{24}. Aged 28, was normal in height, with normal secondary sexual development—his intelligence was probably within normal limits. His visual fields were tested by confrontation and although the results are not very reliable due to poor fixation, there appeared to be nasal hemianopsia in the right eye, and possibly temporal hemianopsia in the left eye. These findings were kindly confirmed by Mr. K. A. Harwood, a visiting optician. No other general or local abnormality was detected on examination except a divergent strabismus in the left eye.

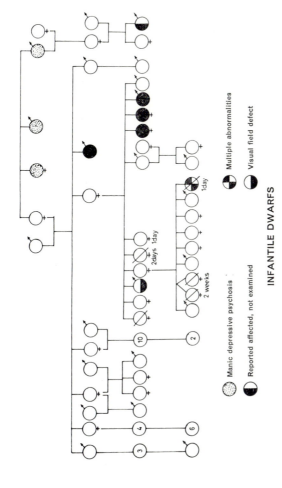

INFANTILE DWARFS

Fig. 4.12. Pedigree of infantile dwarfs.

97

III$_{25}$. Normal height, normal intelligence and secondary sexual development. She had borne nine children, none of whom appeared to be affected.

III$_{28}$. He had left the island, but was reliably reported to have normal mental and physical development. He had two children.

III$_{31}$. She was 14 years old when examined, yet she had the appearance of a girl of 8 or 9. The intelligence was apparently normal. There was no evidence of any secondary sexual development. The bone age was delayed 3–4 years. The limbs and body appeared to be well proportioned, the detailed measurements have been given below.

III$_{33}$. Aged 13$\frac{1}{2}$, was considered a normal prepubertal boy. The bone age was not delayed.

III$_{32}$. Aged 9, was difficult to assess; the only definite abnormality detected was a bone age of 6 years. Standing height was 120 cm with crown pubis measurement 60 cm and pubis heel 56 cm, the arm span was 120 cm and head circumference 52 cm.

II$_{9}$. Aged 53, is a man of somewhat childish personality and reduced intelligence who is considered a local joke. He is employed in simple manual jobs. The standing height was 143 cm, with a crown pubis measurement of 68 cm and pubis heel of 75 cm and the arm span was 160 cm. The weight was 98 lb. He shaved infrequently, had not married nor produced children. The blood pressure was 120/70.

No abnormality of the pituitary fossa was seen in the X-rays of any individual.

In summary, the findings are of individuals with reduced height and bone age, who have a childish appearance and no secondary sexual development. The body weight was also reduced. These features are typical of pituitary deficiency (Hubble, 1965); however, no patient showed reduced skin pigmentation, and Dr. E. Ross considered that the urinary steroid assay made this diagnosis unlikely. When this family was discussed with Dr. J. M. Tanner he suggested that it may be difficult to exclude pituitary deficiency. Familial pituitary dwarfism occurs in mice (Francis, 1944), and has been claimed in man by Zychowicz (1964). The same family was later reported by Trygstad and Seip (1964) who showed response to human growth hormone. Subsequent reports have established that familial dwarfism may be due to pure growth hormone deficiency (Rimoin *et al.*, 1966), which would perhaps seem the likeliest diagnosis in the present family.

Congenitally stiff elbow

Limitation of the range of elbow movements was noticed when measuring the circumference of the upper arm to obtain a correction factor for blood pressure estimation. The limitation was symmetrical, permitted movement between 30–150 degrees and was said to have been present from infancy although it was symptomless. Due to lack of time and the mildness of the anomaly the families were not investigated.

No other joints were affected, and no other abnormalities were discovered except for short, broad finger nails in two people. The possibility of nail patella syndrome was considered, but Dr. J. Renwick, who was consulted, said that it was not. It may represent another form of the dominant congenital rigidity of the elbows and knees described by Pasma and Wildervanck (1956).

Gross unilateral familial genu valgum

Six individuals were seen with a type of genu valgum that has not been described previously. It differs from the usual variety in that it is predominantly unilateral, the onset is at puberty and there is more than 9 cm separating the medial malleoli, which is the limit of the idiopathic variety (Morley, 1957). No other abnormality was found in any system. The intelligence and sexual development were normal.

The trunk was slightly shorter than expected but detailed measurements (below) showed no gross disproportion. People felt somewhat ashamed of their deformity so that it was only possible to obtain measurements on the least severely affected, who had 24, 19 and 13 cm of separation between the medial malleoli.

The patients and their families stated that no abnormality was present until puberty when the knee was injured and the resulting deformity slowly progressed thereafter. The measurements given above were taken on male patients aged 50, 34 and 17 years respectively, which would support the stated slow progression of this deformity, which became a marked disadvantage by middle age because it looked unsightly, prevented normal walking and developed secondary arthritic change.

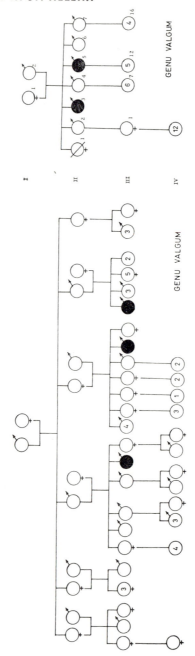

Fig. 4.13. Pedigree of gross unilateral genu valgum.

Anticipating the possibility that this deformity might be a form of X-linked rickets, serum and urine samples were collected from the affected individuals and their relatives and sent to Dr. E. Frater at Cape Town University and to Professor C. E. Dent of University College Hospital who kindly performed serum phosphorus, calcium and serum alkaline phosphatase estimations, and found them within normal limits.

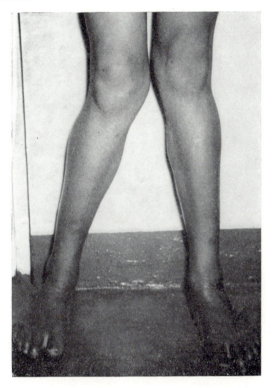

FIG. 4.14. Gross unilateral genu valgum.

Radiological investigation did not show any obvious specific abnormality, although in the older individuals secondary arthritic change with some destruction of the lateral tibial condyle was apparent. Sir Herbert Seddon, who was consulted, stated that the bone texture was about normal, and that the disease would appear different from the symmetrical bone-growth disorder described by Mary Morley (1956). As there was no evidence of bone-softening disease, such as tubular rickets, he suggested that it may be an unknown disorder like Blount's tibia vara (1937).

The pedigrees obtained do not clearly establish the type of inheritance, although rendering a dominant gene hypothesis

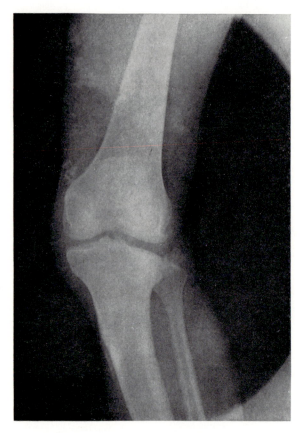

Fig. 4.15. X-ray from unilateral genu valgum.

unlikely. Although all affected individuals were male, the male to male transmission is evidence that this is not an X-linked trait, but this rule is not absolute since there was much inbreeding and incest in these families. The disorder may be determined by an autosomal recessive gene; the obvious failure to obtain the expected segregation ratio may be because V_{4-7} and VII_{3-9} (Fig. 4.13) have

102

not yet reached puberty. An additive gene trait offers another possible interpretation.

TABLE 17. BODY MEASUREMENTS (IN CENTIMETRES) ON PATIENTS WITH GENU VALGUM

	Normal	B.P.		R.T.		H.A.	
Sex	F	M		M		M	
Age	55	34		17		51	
Head circumference	51	54		54		53	
Arm span	158	175		164		183	
Height	168	167		161		171	
Crown pubis	73	78		76		76	
Pubis heel	85	89		85		95	
Dist. between medial malleoli	0	19		13		24	
Hand span	20	19	19	19		22	23
Hand length	18	18·5	18·5	18		21	21·5
Mid finger length				8			
Acromion-olecranon	32	37	34	34		36	37
Olecranon-ulna styloid	25	27	27	25		28·5	28
Clavicle	15			11			
Sternum	14			13			
Inter spinous	22			24			
A.s.i.s.—tib. tub.	52	57	57	54	53		
A.s.i.s.—med. mal.				86	85		
Gtr. troc.—Add. tub.	35	41	40	38	37		
Tib. tub.—med. mal.		35	35	34	34	38·5	40
Foot width	10			10	9·5		
Foot length	25·5			23	22		
Big toe	7·5			7	7		
Serum calcium mg per 100 ml		10·0		10·2		10·0	
Serum phosphorus mg per 100 ml		5·2		4·2		3·6	
Serum alkaline phosphatase u.		13·5		12·0		10·0	

Harelip and cleft palate

The criteria of acceptance were discontinuity of tissue, or the presence of an operation scar.[1]

The following groups of patients were seen:

---

[1] Cases of any disease discovered by an operation scar were accepted if the nature of the operation could be confirmed from the hospital records. Examples of pyloric stenosis, thyroglossal cyst and synotosis of the toes were detected in this way.

(1) Harelip alone. Three females, one male—they form part of a large pedigree resembling an irregular autosomal dominant inheritance.

(2) Harelip and cleft palate. Two females and one male (sporadic).

(3) Cleft palate alone. Four females and two males not obviously related.

Microfilms of harelip and cleft palate, such as rotation and crowding of the front teeth or absent incisors (Fukuhara & Saito, 1963) were not looked for, although some borderline examples were seen. One boy and one girl had a deep furrow at the apex of their palates; another had the appearance of a repair scar that has not healed well, and another had two small grooves about $\frac{1}{8}$ in. in depth at the base of both nares that looked like the beginning of a bilateral harelip. None of these were included in Table 13.

Congenital heart disease

Only patients with gross evidence of cardiac abnormality associated with effort intolerance, in the absence of other causes, were listed as examples of congenital heart disease. An additional definite example was not included, since the patient was listed under a different heading; and there were several patients seen who were not included since they did not fulfil at least three of the following criteria of gross cardiac abnormality:

(1) A grade III or IV murmur.

(2) Any murmur accompanied by a thrill.

(3) Radiological evidence of cardiac enlargement.

(4) Abnormal electrocardiogram.

(5) Central cyanosis.

In addition, two cousins with right bundle branch block were not included since the disease may have been acquired, and, furthermore, they only satisfied the fourth criterion above.

Clearly some genuine cases may have been overlooked, but at least those accepted give a reliable minimum estimate of frequency. The total number of people with the diagnosis of probable or definite congenital heart disease was ten, and although the type of anomaly differed from one to the other, many segregated in sibships or were related as cousins. Families showing several unrelated types of congenital heart disease among close relatives were reported by Carleton *et al.* (1958).

Congenital hyperbilirubinaemia

Four adults were seen who gave a history of jaundice present since birth with no fluctuation in intensity. They never passed pale stools nor dark urine, they had no pruritus or malaena or haematemasis. No patient gave a past history of any major illness except a lady of 52 (II_{12}) who had uncontrolled thyrotoxicosis which was treated with propylthiouracil 25 mg t.d.s. for one year when she developed aplastic anaemia and died. A 32-year-old man (III_{21}) reported that on two occasions he had passed red urine after drinking an excess of spirits, but he resisted attempts to persuade him to repeat the experiment. He also complained of occasional rashes on the hands and feet.

Physical examination of each patient revealed yellow skin over the entire body, most noticeable in the mucous membranes and sclera. No patient had an enlarged liver, spleen or lymph glands, no patient had spider naevi, palmar erythema or abdominal mass. II_{12} had a diffuse goitre and the signs of thyrotoxicosis. III_{21} had small areas of eczema on the exposed parts of the feet and backs of the hands, but no other abnormality was detected on any patient.

The results of laboratory investigations are given in Table 18.

With the possible exception of II_{12}, whose clinical picture was confused by thyrotoxicosis and the propylthiouracil which she had been given, no patient showed evidence of increased haemolysis. The haemoglobin was normal, there was no increased reticulocyte count, no increase in red cell fragility, and no excess of urinary urobilinogen.

The patient who gave a history of passing red urine produced a urine that turned black on standing, following the administration of 4 g salicylamide; but the pigment was not identified. It has been shown by C^{14} studies (Alpen *et al.*, 1951) that 1% of ingested salicylate is excreted as gentisic acid and when Professor G. Rimmington was consulted, he recalled seeing patients who passed black urine after salicylate poisoning. To exclude the possibility of porphyria cutanea tarda in this patient, urine and stool specimens were sent to Dr. E. B. Dowdle at Groot Schuur Hospital, Cape Town, and Professor G. Rimmington, who both kindly undertook porphyrin analysis and found normal quantities to be present.

TABLE 18. RESULTS OF INVESTIGATION OF PATIENTS WITH CONGENITAL HYPERBILIRUBINAEMIA

Pedigree No.	II₁₂	II₁₅	II₁₉	III₂₈	Control
Age	52	64	54	32	28
Clinical appearance	Thyrotoxicosis Deep jaundice	Deep jaundice	Deep jaundice	Eczema Deep jaundice	Normal
Age onset	Birth	Birth	Birth	Birth	–
Fluctuation	–	–	–	–	–
Direct serum bilirubin (mg per 100 ml)	0·6	–	0·7	0·7	0·45
Indirect serum bilirubin (mg per 100 ml)	12	–	15	15	1·65
Icteric index	–	–	48	50	14
Urine bilirubin	– ve	– ve	– ve	– ve	– ve
Urine urobilinogen (titre)*	+ 1/160	–	+ 1/10	+ 1/10	+ 1/10
Serum cholesterol (mg)*	–	–	–	180	–
Haemoglobin (%)*	56	–	95	92	104
Reticulocytes (%)	0·6	–	0·7	0·7	0·5
Alkaline phosphatase (Bodansky) units	–	–	–	19	–
Thymol turbidity (units)	–	–	–	0·5	–
R.B.C. fragility (% saline solution)	0·33–0·21	–	0·42–0·36	0·45–0·36	0·42–0·36
Red cells (millions per mm³)	3·7	–	–	4·2	–
Mean corpuscular volume (c)	–	–	76	90	–
Urinary glucuronide (mg per 4 hr)	–	–	–	2420	2190

* Estimated in photoelectric colorimeter.

There are three groups of patients described showing an increased unconjugated serum bilirubin without any increase of conjugated bilirubin. In so-called Gilbert's syndrome (Gilbert *et al.*, 1907), there is a fluctuating level of unconjugated bilirubin rarely rising above 4 mg per 100 ml, occurring in patients who are otherwise healthy and who tend to transmit the trait to their offspring (Dameshek & Singer, 1941; Alwall *et al.*, 1946; Baroody & Shugart, 1956). In Crigler–Najjar's syndrome (Crigler & Najjar, 1952), the serum bilirubin level generally varies between 13 and 48 mg per 100 ml serum, all or almost all of which is unconjugated, and which shows autosomal recessive type of inheritance (Childs *et al.*, 1959; Szabö & Ebrey, 1963), which authors were able to demonstrate partial impairment of glucuronide formation in the heterozygotes, although others have not been so successful (Billing, 1964). Out of the 19 people so far described with this syndrome, all but 5 have died in childhood of severe brain damage, but these 5 are quite healthy apart from the jaundice (Schmid, 1966). The third group of patients (only 9) differs from the others, having an intermediate level of bilirubin, 6–20 mg per 100 ml serum, that develops from 1 to 30 years of age usually producing no brain damage, and there is usually diminished excretion of urinary glucuronide (Arias, 1962).

This St. Helenian family seems distinct from Gilbert's syndrome on the basis of bilirubin level and the absence of dominant inheritance. The normal results of the salicylamide conjugation test were checked by Dr. Barbara Billing, who had kindly given instruction on the techniques required in using the test. Since it is probably measuring a direct gene product it seems reasonable to separate the hyperbilirubinaemias into those with a low and those with a high level of serum bilirubin and divide the latter into those that can conjugate salicylamide and those that cannot (presumably they lack another enzyme that is part of an alternate pathway). This is preferable to a clinical separation based on the amount of brain damage, which is a remote gene effect, being modified by the time of onset and many other factors.

The pedigree is consistent with autosomal recessive inheritance. None of the affected individuals were known to be the offspring of first-cousin matings, but they were not known not to be, and they certainly shared many common ancestors since they lived in a small, isolated part of the island. It is concluded that this family perhaps represents a new mutation.

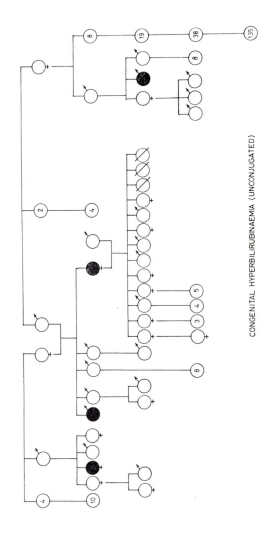

CONGENITAL HYPERBILIRUBINAEMIA (UNCONJUGATED)

Fig. 4.16. Pedigree of congenital hyperbilirubinaemia.

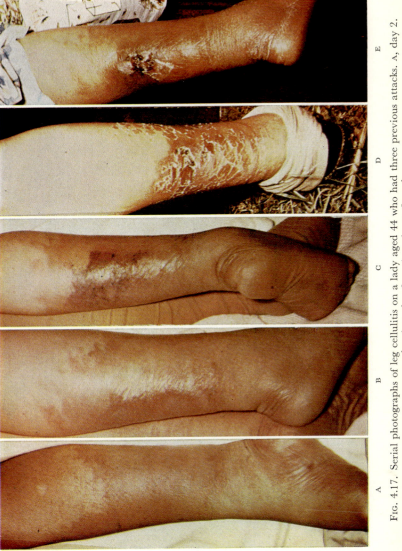

Fig. 4.17. Serial photographs of leg cellulitis on a lady aged 44 who had three previous attacks. A, day 2. B, day 4. C, day 6. D, 4 weeks. E, 6 weeks.

The ingestion of at least one plant, *Senecio* (Schoental, 1960), may cause jaundice, but it is hardly likely to be the cause of unfluctuating jaundice present since birth, as in this family.

Familial St. Helenian fever

Fifteen people were seen during several attacks of an acute febrile illness characterized by an unusual leg cellulitis. This disease had not been previously recognized on St. Helena and the islanders had no name for it. Photographs and case histories were discussed by Dr. P. J. Hare of the University College Hospital, who was of the opinion that the disease is not seen in this country, and when it was demonstrated at a meeting of the Royal Society of Tropical Medicine in 1963 it was agreed that it did not resemble any known tropical disease.

The leg cellulitis was curious in that it was always unilateral, it was recurrent and it had reticulate borders that did not spread up the leg or down onto the foot. Sometimes the cellulitis abated after the third day, but more often, numerous small blisters 1–5 mm in diameter appeared on the back of the ankle; they either resolved or coalesced to form larger blisters 2–17 mm in diameter which would burst, dry, scab and slowly separate during the next 6 weeks, leaving no scar. The progression of the cellulitis is shown in Fig. 4.17 A–E.

Because the illness had an acute febrile onset with rigor, marked systemic disturbance and a polymorphonuclear leucocytosis, an acute infection was considered a likely diagnosis. Accordingly, efforts were made to identify the cause. The streptococcus may cause recurrent leg cellulitis (Cowan & Alexander, 1961) but it was not considered responsible in this case, because the infection did not spread, antibiotics and sulphonamides failed to influence the course of the disease, and because blood and blister fluid were sterile after aerobic and anaerobic culture. For this reason a viral aetiology was entertained. A virus does not usually cause cellulitis, but Tatlock (1947) reported it to be the cause of "pretibial fever", which was characterized by epidemics of recurrent leg cellulitis (Bowdoin, 1942; Daniels & Grennan, 1943).

An attempt was made to inject day-old albino inbred mice by inoculating them intracerebrally with blister fluid, but Dr. J. Gear of the Poliomyelitis Institute of Johannesburg was unable to isolate any virus. Mr. A. Loveridge, former curator of reptiles of the

109

Museum of Comparative Zoology, Harvard, kindly searched Sandy Bay for a likely vector, as most of the cases occurred in Sandy Bay. He discovered two species of black fly (Simuliidae), one of which was hitherto undescribed (Crosskey, 1965). Dr. Crosskey did not think it likely that the black fly would bite man, and few patients gave a recent past history of a bite, so the aetiology was unsettled.

When this illness was reported (Shine, 1964) it was not realized that familial Mediterranean fever, which is known to be determined by a recessive gene in the homozygous state (Sohar et al., 1961), bears some resemblance to it in that both have an acute febrile onset, both are recurrent and both have an erysipelas-like leg cellulitis. Photographs were sent to Professor Heller who has described familial Mediterranean fever (Heller, 1962): he thought that the diseases were not parallel, although agreed that there were certain features in common.

When the affected individuals were assembled according to their relationship to each other, the resulting pedigree (Fig. 4.18) suggested that the illness may be inherited. Bearing in mind that the unaffected people on the pedigree were those who were not observed during an attack, it is possible that it might be more suggestive of a segregating trait if the families could be followed for many years.

It is considered possible that this illness is the result of a genetically-determined susceptibility to an infection.

Ichthyosiform erythrodermia

The recessive form of ichthyosis is characterized by the presence of dry, scaling skin distributed on the face, trunk and limbs including the popliteal and antecubital fossae (Butterworth & Strean, 1962). These features were found on three patients, two of whom were the offspring of a half-brother–sister mating; and were reported by reliable witnesses to be present in I_5 and II_3 who were not examined and so not included in the list of disorders, although the features were so well described that the diagnosis cannot be doubted. The facial appearance of one of them was so striking that the islanders had composed a ditty to it.

The combination of the pedigree (Fig. 4.19) and typical phenotype provides good evidence of recessive determination. Several people were seen with a skin lesion resembling ichthyosis, often associated with varicose ulceration, but this type of ichthyosis was

110

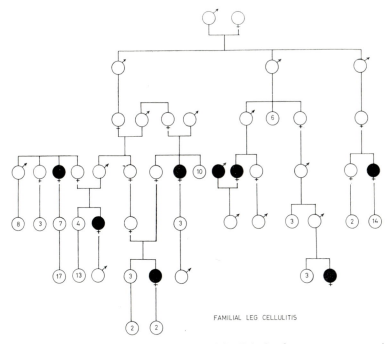

FAMILIAL LEG CELLULITIS

FIG. 4.18. Pedigree of familial St. Helenian fever.

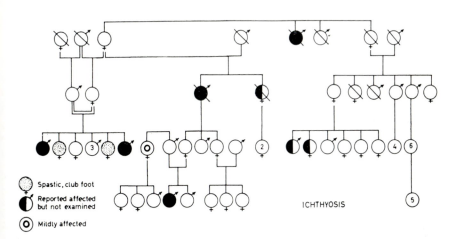

Spastic, club foot

Reported affected but not examined

Mildly affected

ICHTHYOSIS

FIG. 4.19. Pedigree of ichthyosiform erythrodermia.

excluded as it was clearly secondary to the local disease, and did not segregate in families.

Mental deficiency without other abnormalities

No intelligence tests were locally available; Raven's progressive matrices were ordered but did not arrive, hence the diagnosis rested on clinical assessment supplemented by school teacners' and employment officers' reports. No child was diagnosed as defective unless the school teacher considered him ineducable, and no adult was diagnosed as defective unless incapable of normal employment. Hence the estimate of thirty-four individuals with an I.Q. below 60 is somewhat crude.

On the suggestion of Professor C. E. Dent, the urine of affected individuals was routinely tested for cystine, sugars, protein, calcium, acetone, phenylpyruvic acid, keto acids and indican, but no specific abnormality was found. Sixteen individuals were not tested as specimens could not be obtained.

Microcephaly

Penrose (1956) suggested reserving the diagnosis of true micro-cephaly for a recessively determined disease characterized by short stature, reduced intelligence and a small head with low cephalic index, provided it is not secondary to craniosynostosis or part of a general disorder such as mongolism. Two individuals seen fit this definition; they had low intelligence, were short, but in proportion, and the head circumference was between two and three standard deviations below normal for their age and sex. The patients were not related and were not known to be the progeny of first-cousin marriages. Both have normal parents and both have two normal sibs.

Microphthalmos

Three sporadic cases were seen, one with bilateral microphthalmos and mental deficiency, the other two with unilateral microphthalmos and normal intelligence. There was, however, some reduced vision in the normal-sized eye. Skull X-rays revealed no evidence of toxoplasmosis and there was no history or evidence of trauma.

No one had any other bony or congenital anomaly or any of the associated syndromes mentioned by Falls (1966).

Little family history could be elicited, but apparently the three people were unrelated and represent sporadic cases which have been found more frequently than those with obvious familial segregation (Sjørgren & Larsson, 1949). Apart from the dominant and sporadic types mentioned by these authors, Holst (1952) described a recessive variety that also occurs in pigeons.

Mongolism

There were five sporadic unrelated mongols seen; a male, aged 2 years, and another male aged 23 years, and three females, aged 3, 11 and 25 years. An increased frequency of mongolism might be anticipated in an inbred community if recessive genes predispose to non-disjunction.

Multiple neurofibromatosis

The only unequivocal example of this disease was a 43-year-old male who developed a sarcoma of the right thigh, which was removed. A few people were found with scattered papillomata and one or two café-au-lait spots, but these were not included in the list of genetical disorders, since Crowe et al. (1956), in an extensive investigation, found that six café-au-lait spots were necessary for the diagnosis. These authors found sarcoma in 2% of their cases.

This patient had one affected and three unaffected sibs, and one affected parent.

Muscular dystrophy

Two boys aged 16 and 13 and a girl aged 3 years presented me with an unusual muscle-wasting disease. The older boy, V_{15}, had asymmetrical wasting that was almost entirely confined to the right shoulder girdle. He had lost all movement of the right arm except rotation, which he was able to perform against gentle resistance. The biceps, triceps and supinator reflexes were absent in the right arm but elsewhere the reflexes were normal. There was no fasciculation. Although there was no gross weakness or wasting of other muscle groups there was a generalized slight loss of muscle power. There was no visual defect and no ptosis.

The age of onset and rate of progression are uncertain as some informants claimed that the arm was weak since birth and others since the age of 5. V_4 had a history of global weakness starting at the age of 6, leaving him with a very slowly progressive weakness and deformity in the upper limbs and slight wasting and weakness of

113

the lower legs and feet. The power was generally greater than the degree of wasting suggests, and it increased proximally. Tone and co-ordination were normal, reflexes were brisk and bilaterally symmetrical. There was no fasciculation. No abnormality in sensation was detected in either patient. V_9 was symptomless and had no complaint, but on examination was found to have winging on the scapulae.

Until recently, the muscular dystrophies were an ill-defined heterogeneous group, but recent research (Stevenson, 1953; Walton

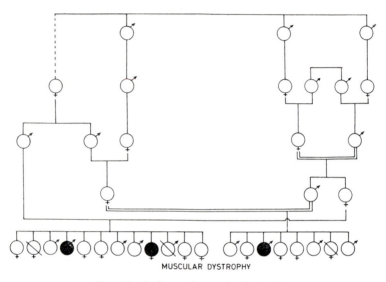

MUSCULAR DYSTROPHY

Fig. 4.20. Pedigree of muscular dystrophy.

& Nattrass, 1954; Morton *et al.*, 1963) has clarified the classification into five reasonably distinct groups, although all authors admit other types that do not fit into any of the major categories. The present family comes closest to the group with late onset distal myopathy described by Gowers (1902) and Welander (1951), although differing in showing an early age of onset and predominantly unilateral disease in at least one individual. While Walton (1963) recognizes unilateral variants and Bourrat *et al.* (1963) have described varieties intermediate between amyotrophic lateral sclerosis and scapulo humeral myopathy, the present family would seem unclassifiable (Fig. 4.20).

114

Ocular nystagmus

One of the five adults examined with idiopathic nystagmus had a high-pitched squeaky voice that would have been considered a chance association but for the report from the neighbours that several of her sibs had similar eyes and a similar voice: "when there was a room of them together they sounded more like mice than Christians".

Two sibs were examined who were normal, but the girl's mother had nystagmus and, at times, her voice squeaked as though she were controlling it by conscious effort, relaxing at odd moments. Indirect laryngoscopy did not allow adequate visualization of the vocal cords. There was no stridor and the patients were able to cough explosively. Neither mother nor daughter had the appearance of *cri du chat* syndrome.

It is suggested that a medullary lesion might account for this association since the larynx is innervated by the vagus which comes from the nucleus ambiguous, and lesions in this region may cause nystagmus, but in this instance, it may be a chance or environmentally determined anomaly. Plott (1964) has described a sibship of seven, of whom three brothers had a congenital laryngeal abductor paralysis which he attributes to a genetically determined lesion in the nucleus ambiguous. Nystagmus, without other anomaly, showed sex-linked recessive inheritance in one family.

Palilalia and echolalia

Two brothers presented an unusual and similar clinical picture. One had echolalic speech and appeared to have a low intelligence, and the other had palilalic speech and almost normal intelligence. A tape recording was taken of his speech, and an example of his handwriting was obtained (Fig. 4.21) which shows a suggestion of the same tendency. According to their sister, they had a third brother who had the same physical appearance, a low intelligence and echolalic speech. Their physical appearance was unusual and considered grotesque by many islanders. The head was elongated from lambda to chin and the mandible was small. The tongue was not unduly large, but slightly protuberant. They both preferred to adopt a stooping posture with the head, hips and knees flexed, although they were able to stand erect (Fig. 4.22). The gait was shuffling and the movements staccato.

It was hard to assess their intelligence because their difficulty in communication had prevented much social contact. Examination

115

of the central nervous system revealed no gross abnormality apart from the speech defect. The cranial nerves appeared normal, muscle power, tone and co-ordination were normal, there were no involuntary movements and no abnormality of sensation. The reflexes were very brisk but bilaterally equal, and there was no

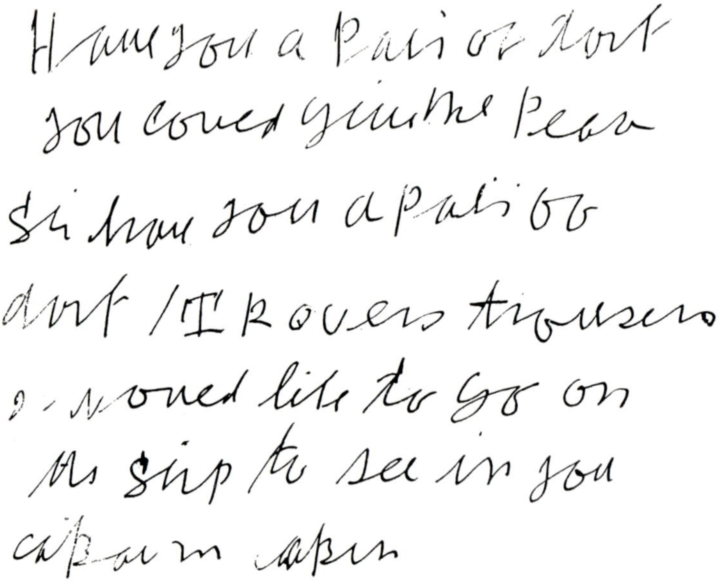

HAVE YOU A PAIR OF DARK YOU COULD GIVE ME PLEASE SIR HAVE YOU A PAIR OF DARK TROUERS TROUSERS I WOULD LIKE TO GO ON THE SHIP TO SEE IN YOU CABAIN CABIN.

FIG. 4.21. Handwriting of patient with palilalia.

clonus, up-going toes or other evidence of upper motor neurone lesion.

An electrocardiogram showed no abnormality; an X-ray of the abbreviated skeleton showed an elongated skull but no other abnormality. The urine was tested for protein, sugars, acetone, cystine, calcium, phenylpyruvic acid, keto acids and indican, and no abnormality was found.

No history of trauma, encephalitis or severe illness was elicited, and close relatives stated that the illness has been present since infancy.

While echolalia and palilalia are considered to have a different aetiology (Brain, 1961), the presence of three brothers with similar rare defects is suggestive evidence of a common aetiology, although not necessarily genetical. None the less, a genetical aetiology must

FIG. 4.22. Patient with palilalia.

be a possible explanation of any familial defect of early onset for which there is no obvious environmental cause, even though this has not been reported previously.

Pterygium

Pterygium was seen in ninety-two adults and in a few school children. This is certainly an underestimate of the frequency as only gross lesions were detected. It is thought that this may be a genetic disease because the environmental factors to which it is attributed, namely dust, wind and dry air, are not found on St. Helena, or certainly less so than in Britain. Pterygium was found with high frequency in the isolated inbred fishing populations of the Åland islands (Forsius & Eriksson, 1962).

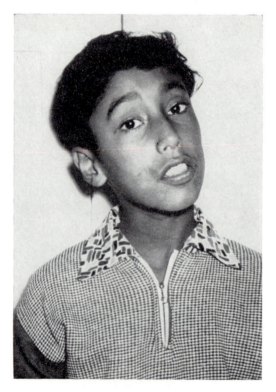

FIG. 4.23. Patient with ocular torticollis.

Ptosis

This common disorder caused little disadvantage in the majority of patients, but one little boy had torticollis that was tentatively attributed to a visual defect caused by a bilateral ptosis. He had a full range of head and neck movements, but held his head on the side (Fig. 4.23) because it appeared to confer some visual advantage. This was most noticeable when he was crossing the open country.

He also had a partial coloboma of the optic nerve and nystagmus and it may be that lesion rather than the ptosis that caused the torticollis, although torticollis secondary to ptosis has been reported (Guy, 1942).

Spasticity

Apart from six spastics who showed no unusual features, there were four individuals showing the following combination of anomalies: moderate mental deficiency (the intelligence quotient was judged to be about sixty), bilateral primary optic atrophy and in addition, two of them had pes cavus. The two without pes cavus had two brothers with pes cavus and ichthyosiform erythrodermia (see Fig. 4.19) which is interesting in view of Sjørgren and Larsson's (1957) report of a syndrome characterized by spasticity, mental deficiency and ichthyosiform erythrodermia, retinal degeneration and speech defect.

Thrombocytopenic purpura

This is the presumptive but unsubstantiated diagnosis made to explain the clinical appearance of purpura on large numbers of islanders. Small haemorrhages, 1–2 mm in diameter, were evenly distributed on the trunk, front and back, and sometimes on the limbs and face, diminishing in frequency distally.

The history of the appearance of the spots was difficult to obtain as most islanders alleged that they were due to flea bites, but after a few months on the island the author had acquired sufficient personal experience to refute this diagnosis with confidence. One islander, a reliable witness, reported that she awoke one morning to find a widespread rash coincident with the onset of menstruation; she was seen the following day when purpuric spots were noted on the face, neck, upper arms and trunk, but she was otherwise free of symptoms or abnormal physical signs. She was given no treatment and when examined one week later the spots had disappeared. Another 14-year-old girl was examined at a morning surgery where she complained of non-specific pain in the chest for 2 days followed by spots on the chest for 1 day. She said that the spots had suddenly appeared and admitted that she had noticed the same appearance 1 month previously just before the onset of menstruation. On examination she was found to have a temperature of 100° and purpuric spots evenly scattered over the front and back of the chest with occasional spots on the face, neck, abdomen and upper arms.

More often no history of onset could be obtained and it was alleged that the spots were always present.

Patients did not complain of haemorrhagic symptoms except those who overlapped with the family with Christmas disease (Fig. 4.24). It was thought probable that the disorder generally became milder with increasing age since the majority of affected individuals seen were school children and young adults. There were no apparent sequelae to this disorder with the possible exception of a lady of 34, who having had purpura for 5 years developed a polyarthritis and I have recently heard that she died of leukaemia.

Simple laboratory tests usually gave normal results; only three patients had a positive tourniquet test. Platelet examination appeared to be normal, but even in skilled hands this investigation can be misleading.

Von Willebrand's disease apart, the inherited thrombocytopenias are a confused and heterogeneous group (Ata *et al.*, 1965; Quick & Hussey, 1963; Hardisty *et al.*, 1964) with a large recessive component (Hardisty *et al.*, 1964; Pittman & Gaham, 1964) and as much as 10% parental consanguinity (Larrieu *et al.*, 1961). Since the pedigree and laboratory information on these families is poor, they cannot be categorized beyond saying that they are clearly not examples of Aldrich's syndrome (Aldrich *et al.*, 1954), which is usually lethal in childhood.

Syndrome of low rasping voice and low intelligence

Seven individuals with a low intelligence and a strange voice were examined. Each individual was considered educationally subnormal although they were all able to remain at school and later held simple but everyday jobs. The unusual feature common to both sexes, children as well as adults, was a deeply pitched, rasping voice.

No other physical abnormality was detected on examination. Indirect laryngoscopy did not reveal the cause of the anomaly, because I was unable to obtain a good view of the cords.

All affected individuals were related and four of them were offspring of a cousin marriage, suggesting the possibility that the anomaly may be the effect of a single gene. The segregation ratio does not differ from that expected on this hypothesis (Table 19). Lejeune (1965) has seen patients with low intelligence, a prominent bridge of the nose and deep voice in association with trisomy of

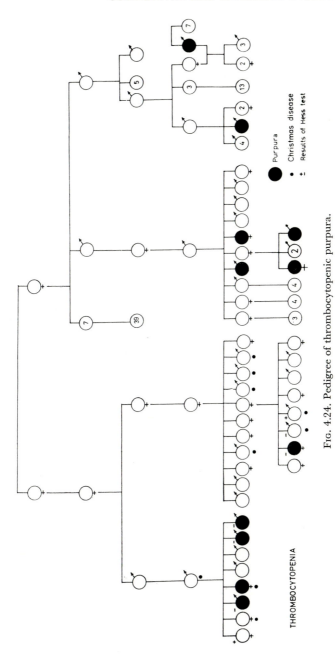

Purpura

Christmas disease •

Results of Hess test ±

THROMBOCYTOPENIA

Fig. 4.24. Pedigree of thrombocytopenic purpura.

TABLE 19. HOGBEN'S (1931) METHOD OF TESTING A RECESSIVE HYPOTHESIS ON SIBSHIPS CONTAINING CASES WITH A LOW RASPING VOICE AND A LOW INTELLIGENCE

Number in sibship (s)	Number of sibships (n_s)	Affected		Variance of expected number ($n_s\sigma_s^{-2}$)
		Observed	Expected	
3	1	1	1·2973	0·26297
4	2	1	2·9256	0·84010
8	1	3	2·2225	1·1724
11	1	1	2·871	1·8053
TOTAL		6	9·3164	4·08077

The difference between observed and expected $= 2\cdot32$ is less than two standard deviations $= 4\cdot04$ of the expected number.

the short arm of chromosome No. 5. A recessive syndrome consisting of lipoid infiltration of the skin and larynx gives the skin a mottled, lumpy appearance which these patients did not show, as well as a hoarse voice (reviewed by Gorlin & Pindbury, 1964).

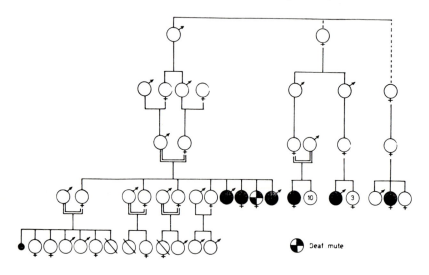

DEEP VOICE AND LOW I.Q.

FIG. 4.25. Pedigree of voice anomaly and low intelligence.

DISCUSSION

Many entities in the foregoing compilation resisted precise diagnosis, some eluded separation into predominantly genetical or predominantly environmental categories and even for those that are clearly genetical the data were often insufficient to establish with certainty whether they were dominant, recessive, sex-linked or additive gene traits. Bearing in mind also the precautions listed at the beginning of this chapter, it seems difficult to make much useful comment. However, the anomalies listed in Table 20, being well defined and unambiguous, can be properly pigeon-holed and compared with other studies.

It is apparent from Table 20 that there is an increased frequency of certain traits that are rare in larger populations. It is also apparent from the phenotypes alone that many of them can be attributed

TABLE 20. FREQUENCY OF TRAITS IN LARGER POPULATIONS

Trait	Prevalence in St. Helena	Prevalence in panmixia	Country	Source
Generalized albinism (oculocutaneous)	5/4259	5/57,100	N. Ireland	Froggatt (1960)
Deaf mutes with retinitis pigmentosa	1–4/4259	1/33,000	Sweden	Hallgren (1959)
Deaf mutes	8/4259	8/34,000	N. Ireland	Stevenson and Cheeseman (1956)
Christmas disease	3–21/4259	3/480,000	Sweden	Ramgren (1962), Ikkala (1960)
Ichthyosiform erythrodermia	3/4259	3/900,000	England	Wells and Kerr (1966)
Mongolism	5/4259	5/19,890	Denmark	Oster (1953)
Microcephaly	2/4259	2/500,000	Holland	Van der Bosch (1959)
Microphthalmia	3/4259	3/120,000	Sweden	Sjørgren and Larsson (1949)
Crigler–Najjar type hyperbilirubinaemia	4/4259	28 in world literature		Schmid (1966)
Lithopaedion	1/4259	11 in world literature		Lord (1956)
Brachydactyly	32/4259	1/1000	U.S.A.	Temtamy (1965)

123

to autosomal recessive genes. Furthermore, some of the unique syndromes mentioned previously seem to be similarly determined. As an increased frequency of rare recessive traits, and the revelation of some that were hitherto unknown are the foremost expected consequences of inbreeding, it is natural to ask whether the observed pattern of disease is, in fact, due to inbreeding. This is considered in the subsequent chapter.

CHAPTER 5

INBREEDING

Thou shalt not let thy cattle gender with a diverse kind; thou shalt sow thy
field with mingled seed.
(LEVITICUS, 597 B.C.)

The absolute risk for the individual couple (in the absence of any history of
recessive, or possibly recessive defects in the family) is probably quite small.
(FRASER ROBERTS, 1963)

The total risk (for all genetic abnormalities) is quite appreciable and may be
2 or 3 times as large as that from children from unrelated marriages.
(C. C. LI, 1961)

Inbreeders (i.e., families who habitually inbreed) suffer by outbreeding and
outbreeders by inbreeding.
(C. D. DARLINGTON, 1960)

THESE quotations illustrate the long-standing confusion over the
consequences of inbreeding, which arises because people mean
different things by inbreeding, estimate its effects in different ways,
or assume that local results have universal validity. It usually
means the marriage of near relatives such as first cousins, who share
common grandparents, but it may refer to the marriage of individuals
who have any remote ancestors in common. Due to the difficulty
of obtaining reliable information about remote ancestors, it is usual
to count only grandparents or great-grandparents. If all paths of
descent from all common ancestors are considered, ignoring
mutation, selection and drift, it appears that since we are all ultimately descended from two individuals, everyone has a 1 in 2
probability of genetic identity with everyone else. For example,
if Noah was an albino with no resulting loss of fitness, as suggested
by the description in the Books of Enoch and Jubilee (Sorsby, 1958),
approximately one in four children born today will be albinos.
Since the rediscovery of mendelism there has been no disagreement
over the proposition that the offspring of relatives are more likely
to be homozygous than the offspring of strangers. While this applies
to all loci, because homozygosity for common alleles is quite likely
to occur by chance, the effect of inbreeding is not noticeable. But

125

since a particular individual is unlikely to carry a particular rare gene, he is even less likely to carry two identical rare genes. However, if his parents were first cousins, he has a 1 in 32 probability of homozygosity for any gene carried by either parent regardless of the exact degree of rarity of the gene in the population. As this probability is fixed, it becomes relatively more important as the gene becomes more rare.

In applying theoretical models to a practical situation, two common errors occur. Firstly, it is claimed that inbreeding is not harmful, meaning it caused no profound alteration in the population statistics or it caused no disease in one particular group studied. By the same token, poliomyelitis may be considered not harmful, since, like inbreeding, it is uncommon in most countries and some population groups are resistant to the disease. Secondly, the risk is underestimated when inbreeding is considered to bypass all loci but one. In his interesting discussion on the genetical indications for abortion, Smithels (1966) found cousin marriage to be relatively harmless: "First cousins carry 1 in 32 risk of such homozygotes, and this is so close to the random 1 in 40 chance of defect that it must be accepted."

His argument is based on the unstated assumption that the average person harbours only one deleterious autosomal recessive gene, but if the average number of such genes per person is 5 or more, as is generally believed (Stern, 1960), then the risk of defect in the offspring of a first cousin becomes

$$1 - (1 - 1/32)^5 \doteq 1 \text{ in } 7,$$

which is considerably different from 1 in 40.

St. Helena presents a specific practical problem, namely what is the effect of cousin marriages on the production of congenital and inherited disease? It may be resolved in two ways. The frequency of cousin matings among parents with normal and abnormal offspring may be determined and compared. This method was attempted but it failed because it was often impossible to be sure if a particular couple were cousins or not. Pedigrees were compiled for 1031 matings, but of these only 669 were sufficiently detailed to allow the detection of first-cousin matings and only 497 were detailed enough to permit the detection of not-first-cousin matings. It would have been far easier had it been possible to ask people directly whether their parents were related, but it soon became obvious that this approach was too direct for St. Helenian taste. The second method, which is the one adopted, was to ascertain cousin marriages, select a sib or close relative as a control provided they had contracted a *not*-cousin marriage and compare the offspring of each. The validity of this method hinges upon the impartiality with which cousin and control matings are selected and upon the impartiality with which cousin marriages are contracted. In most human societies, social attitudes to inbreeding are not neutral and hence cousin mating tends not to occur at random. If they are not random, being found more commonly among royalty (as in ancient Egypt and modern Europe), among peasantry (as in modern Egypt and ancient Europe) or among those with lower intelligence (as in Sweden), it is easy to confound inbreeding with other sociological effects.

PREVAILING SOCIAL ATTITUDES IN ST. HELENA

There is some confusion about the nature of prevailing customs, perhaps because the islanders are a community which is the result of recent genetic and cultural hybridization. As far as could be ascertained it seemed that there has been free interbreeding and inbreeding for the past 100 years but, at the same time, it is said that there has been pressure tending to discourage second-cousin mating while permitting first-cousin mating. This is generally attributed to empirical observation: "My mother always says that affliction come if you marry into the family." When asked for the source of this dictum, people often said that farmers had observed first-cousin or incestuous mating to be beneficial to cattle and second-cousin mating to be harmful. This unusual attitude is partly due to confused nomenclature; unlike the author, who throughout this text is using the *Oxford English Dictionary* definitions, the islanders frequently understood a second-cousin mating to mean a first-cousin mating between people who were themselves the children of first cousins. The Anglican Church, to which most people belonged, tended to discourage first-cousin marriage, but some of the other churches did not, and, on the whole, it was considered that there was only slight social disapproval of any cousin mating and that it was weakly enforced and predominantly directed against second-cousin mating.

Incestuous relationships were quite infrequent, producing about

$0 \cdot 1\%$ of the population; only four individuals who were the result of such matings were encountered.

Having contracted a cousin marriage, people would not then attempt to restrict the size of their family even if they considered that the risk of abnormality might be increased; the general attitude was a fatalistic and passive acceptance of the inevitable "if that's your luck, it's got to be"; moreover, few people restricted the size of their family for any reason. Condoms and Dutch caps have become sparingly available in recent years, but they are not widely nor intelligently used and rarely in conjunction with a spermicide. Due to the attitude to illegitimacy, induced abortions occur infrequently (less than 1% of all pregnancies) and infanticide is very rare.

IDENTIFICATION OF COUSIN AND CONTROL MATINGS

(a) *Compilation of pedigrees*

Reasonably accurate birth, marriage and death records for the past 100 years were available, but it would have taken too long to compile pedigrees from this source. As previously mentioned, it was not possible to ask the St. Helenians directly if they were related to their spouse. But they did not mind assisting in the construction of pedigrees provided that I, rather than they, put both sides together, and provided that the local rules of conversation, which were explained in Chapter 1, were respected. Illegitimacy was no problem, as the biological father was invariably known and acknowledged, and once the unfamiliar manner of imparting information was understood it was found that most people were entirely open. The high official illegitimacy figure is based on the legal definition which does not take common-law marriages into account.

Because biological fathers were not always the legal fathers, it was necessary to spend a lot of time checking that an apparent cousin marriage was a real one and that an apparent not-cousin marriage was also real. This was made more difficult because the identifying information on each individual was poorly collected. Sensing the islanders' reluctance to give full Christian names and maiden name, I had tended to accept whatever information they offered readily. It took so long verifying cousin and control matings that it was decided not to attempt to confirm the paternity of the offspring of these matings; instead, the simplifying assumption was

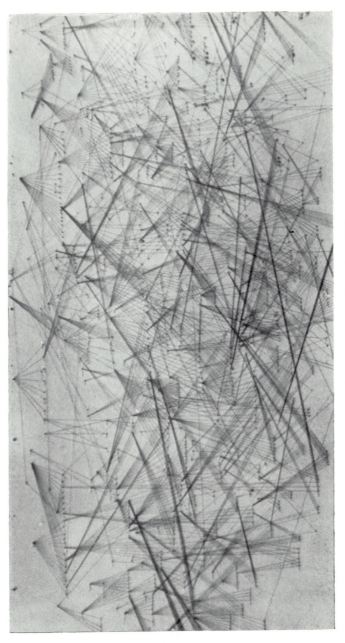

FIG. 5.1. Large complex pedigree from 1780.

made that non-paternity would be as likely to occur in the cousin matings as in the controls.

(b) *Selection of cousin matings*

One thousand and thirty one pedigrees were compiled, some of them extending back to 1780 (Fig. 5.1), although the majority extended only for one or two generations. After leaving the island, each mating represented by at least one living partner was tested for evidence of consanguinity, considering biological relationships only, Because of the difficulty in detecting distant relationships with any degree of reliability, it was decided to limit the investigation to matings between near relatives. It was intended to use increasingly distant relatives until 50 such matings had been collected. The search stopped when examination of matings with a co-efficient of relationship of one-sixteenth or greater produced 51 suitable matings, but of these, 6 were later discarded, 2 because the couple had been married less than 6 months, 3 because the wife was over 40 years of age at the time of marriage and 1 because no control was available.

(c) *Selection of controls*

It was considerably more difficult to establish that someone had not married their first cousin than it was to establish that they had. This was because a single-generation pedigree might be sufficient to prove consanguinity if the parents were known to be full sibs:

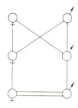

yet a two-generation pedigree with only one missing grandparent

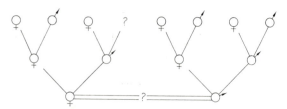

could not be accepted as a not-cousin marriage as the missing person could be one of the grooms' grandfathers. For controls, a sib of the husband and a sib of the wife were chosen, provided that the sib was not married to a cousin. When there were several sibs to choose from, as often happened, the one nearest in age was chosen, and if there were two possibilities, preference was given to one of the same sex, and if there was still ambiguity, preference was given to the one whose Christian name began with the nearest letter. If no sib was available as a control, a first- or second-degree relative was chosen instead, applying the same rules if there was more than one alternative.

If rigid rules had not been applied in the selection of controls, it would have been easy to obtain whatever result was desired since there were usually some members of the sibship with more and some with less detriment among their offspring. It was not considered necessary to choose a control who married an entirely unrelated person, merely one who did not marry a cousin.

Unfortunately, all information had to be memorized as it would seem to them definite evidence of intrusion into their neighbours' or sisters' affairs if I should actually write down in their presence information that they were willing to pass on in conversation. With a few particular friends, however, it was possible to be more direct, and some of them did not mind if the information was written down.

DEFINITIONS

(1) *Stillbirth.* Islanders tended to forget about stillbirths, consequently the entire island apparently has an incidence slightly less than 1% of livebirths. This is also partly because islanders understood stillbirth to refer to children born dead at term, and their answers excluded those born earlier.

(2) *Abortion.* Replies to this question were less reliable than for stillbirths and were soon abandoned.

(3) *Death of offspring* was considered to be fairly reliably reported, but there may have been some underreporting. Since the upper age limit is arbitrary, none was applied, but the age at death was under 10 years in most cases.

(4) *Presence of anomaly.* The examination procedure has been described in the previous chapter. At least, in this category, it is certain that the estimate is reliable. It is not possible that the

examiner could have scrutinized consanguineous and control children differently, because the decision as to which were to be the consanguineous and which were to be the controls was not taken until about one year after leaving St. Helena. Any child reputedly abnormal but not seen, either because of death or just chance, was not included.

(5) *Cohabitation* was taken to commence one year before the birth of the oldest child or at marriage, whichever was the earlier event.

(6) *The coefficient of inbreeding.* F was calculated for each inbred individual from pedigrees using the formula derived from Wright (1922):

$$F = \Sigma \ (1/2)^{n-1} \ (1 + F_A),$$

where n is the number of steps from offspring in a complete loop of descent returning to the offspring; where F_A is the coefficient of inbreeding of the common ancestor, and the summation is taken over all the common ancestors. The coefficient of inbreeding of the common ancestor F_A, is assumed to be zero, although this simplification results in an underestimate of F.

THE SUITABILITY OF CONTROLS

The relationship of the controls to the consanguineous propositi was as follows:

69 sibs	3 nephews
7 first cousins	2 daughters
5 same person, double marriage	1 monozygotic co-twin
3 sons	

For reasons already given, the coefficient of inbreeding of the control group was assumed to be zero.

The impartiality with which the controls were selected was tested by comparing them with the consanguineous propositi for thirteen measured characters, namely age of wife at marriage, age of husband at marriage, duration of marriage, age at time of survey, duration of cohabitation, systolic and diastolic blood pressure, body weight, arm circumference, exercise habit, smoking habit, social class and years of shoe-wearing. No significant differences were found, even using a low level of significance (Table 21). The types of illness found in both groups (Table 22) do not differ, nor is there any difference in occupation (Appendix 4, summarized in Table 23).

TABLE 21. 45 CONSANGUINEOUS AND 90 CONTROL MATINGS COMPARED FOR
CERTAIN STATISTICS

	n	Range	Mean	Significance		
				S.D.	t	p
Age of propositus at time of survey	87	21–76	47·6	16·3	0·53	> 0·5
Age of control at time of survey	87	18–86	46·2	18·3		
Duration of cohabitation						
Consanguineous	88	2–55	24·1	16·2	0·20	> 0·8
Control	87	2–60	23·5	16·1		
Age of wife at marriage						
Consanguineous	44	9–38	20·8	5·3	1·34	> 0·1
Control	88	13–33	19·7	3·9		
Age of husband at marriage						
Consanguineous	43	13–59	24·7	8·5	0·29	> 0·7
Control	88	10–53	24·3	6·6		
Systolic blood pressure						
Consanguineous	59	100–240	146·6	32·2	0·60	> 0·5
Control	59	110–260	150·4	36·0		
Diastolic blood pressure						
Consanguineous	59	65–130	92·8	17·0	0·46	> 0·6
Control	59	65–135	91·4	16·4		
Body weight						
Consanguineous	50	95–185	135·6	21·6	0·86	> 0·4
Control	50	79–210	140·0	29·0		
Arm circumference						
Consanguineous	58	22–34	26·7	2·9	1·70	0·1
Control	58	22–35	27·6	3·0		
Daily exercise (miles)						
Consanguineous	60	0–10	4·6	2·2	1·02	> 0·3
Control	60	0–10	4·6	2·3		
Daily smoking						
Consanguineous	58	0–17	2·7	4·8	0·09	> 0·9
Control	58	0–60	2·6	8·7		
Social class						
Consanguineous	60	1–4	3·6	0·6	0·68	> 0·4
Control	60	1–4	3·6	0·6		
Years wearing shoes						
Consanguineous	60	0–68	11·3	18·2	1·05	> 0·2
Control	60	0–70	15·0	20·4		

TABLE 22. COMPARISON OF GENETICALLY DETERMINED DISEASES FOUND AMONG
THE OFFSPRING OF CONSANGUINEOUS AND CONTROL MATINGS

	Consanguineous	Control
Albinism	1	0
Arthrogryposis and club foot	1	0
Blindness secondary to choroiditis	0	1
Congenital torticollis secondary to ptosis	1	0
Psychotic	1	0
Deaf	1	0
Deaf mute and mentally defective	1	1
Deaf mute (presumed retinitis pigmentosa)	2	0
Deaf mute with retinitis pigmentosa	0	1
Deaf mute with lithopaedion (presumed retinitis pigmentosa)	0	1
Dwarfs with pseudobrachydactyly	2	0
Grand mal epilepsy	0	1
Gross unilateral genu valgum	0	1
Ichthyosiform erythrodermia	2	0
Mental defectives with vocal defect	3	0
Mental defectives	4	0
Muscular dystrophy	0	2
Pulmonary stenosis with reversed shunt	0	2 *
Spastics with pale fundi and low intelligence	4	0
TOTAL	23	10

* Same control used twice.

TABLE 23. COMPARISON OF OCCUPATIONS

	No. who had major illness	
Consanguineous	28/59	$\chi^2 = 0\cdot54$ $0\cdot3 < P < 0\cdot5$
Controls	32/59	

	No. with skilled occupation	
Consanguineous	4/56	$\chi^2 = 1\cdot46$ $0\cdot4 < P < 0\cdot3$
Controls	7/53	

RESULTS

The consequences of inbreeding for the group studied are clear. There is a significantly increased incidence of stillbirth, child death and child defect among the offspring of cousin marriages (Table 24).

Since the groups were comparable in all ways that it has been possible to measure, differing only for their coefficient of inbreeding,

$$F_1 - F_c = 17 \times 1/8 + 125 \times 1/16 + 48 \times 1/32 \div 190 - 0$$
$$= 0 \cdot 0536 \quad \text{(subscripts refer to consanguineous and control respectively)}$$
$$\doteq 1/19$$

it is reasonable to attribute the increased detriment to inbreeding.

Further analysis of this data is in progress using the Morton–Crowe–Muller load theory to quantify the inbreeding effect and segregation analysis to isolate the contribution of major genes.

TABLE 24. OUTCOME OF CONSANGUINEOUS AND CONTROL MATINGS

	No. of livebirths	No. of stillbirths	No. of child deaths	No. of child defects
45 cousin matings	182	8	22	23
90 control matings	415	3	24	10
The difference in frequencies =		0·0349	0·0631	0·1119
Significance of difference X^2		9·2 ($P = < \cdot 01$)	7·2 ($P = < \cdot 01$)	25·9 ($P = < \cdot 001$)

DISCUSSION

By executing this study rather like a double blind control trial, most sources of bias were effectively eliminated. One source that was not tested was that consanguineous parents, perhaps feeling guilty, deliberately concealed stillbirths or infant deaths; however, if this had been so, it would have resulted in an underestimate of the effect of inbreeding.

It is possible that the controls mated assortatively, thereby lessening the difference of F between the two groups, which would result in an underestimate of detriment (Spuhler, 1965). However, there is no reason why cousins should not also mate assortatively (they generally had several possible mates to choose from); and comparison of body weight, blood pressure, arm circumference, shoe-wearing habit and smoking habit has shown no correlation between husband and wife (Shine et al., 1967).

An interesting source of error, common to all studies, is apparent from the deficiency of second- and third-cousin mating in the data, reflecting the great difficulty in making accurate estimates of the inbreeding coefficient. Dahlberg (1948) reasoned that where sibships consisted of B individuals, each individual has two parents who would each have B_1 sibs, who would each have B offspring; then the average person has $2B$ (B_1) first cousins. A more general formula derived by extending Dahlberg's reasoning is,

$$c = 2^{y-1} (B^y - B^{y-1}),$$

where c is the number of y degree relatives, and y is the canonical degree of cousin relationship. It follows that the proportions of third, second, and first cousins are:

$$8(B^4 - B^3) : 4(B^3 - B^2) : 2B(B_1)$$

or

$$4b^2 : 2b : 1,$$

and by substituting 4 for b, which is the average size sibship on St. Helena, the proportions become 64:8:1 or, in absolute numbers, 1536:192:24. Whilst these figures are only crude approximations, they show that it is unlikely that first-cousin marriages are the commonest form of inbreeding and explain why it is always difficult to find entirely unrelated control matings. Since the earliest times most societies have considered inbreeding harmful and have generally prohibited it (Tylor, 1865), although nobody quite knows why. The notable exceptions are the Ptolemies, Incas and Ernadon, who have practised incest without any known resulting detriment. Aberle et al. (1963) have summarized various ideas on the origin of the taboos.

In the last century, the subject was studied by George Darwin (1875) and Bemiss (1858) in humans; by Charles Darwin (1876) in plants, and by Bos (1894) in animals. The theoretical explanation of the ill effects observed was not generally appreciated until the rediscovery of mendelism (Bateson, 1902), when it was used by Garrod (1902) to explain the high frequency of consanguineous marriages among parents of alcaptonurics. In 1951 Sutter and Tabah (also in 1952 and 1957) rediscovered the early American inbreeding studies (Bemiss, 1858; Arner, 1908) and conducted the first modern study of inbreeding in Britanny. This stimulated many workers to investigate the problem (Böök, 1957; Schull & Neel, 1958; Slatis et al., 1958; Dolinar, 1960; Goldschmidt et al., 1963; Tanaka, 1963; Dronamraju & Khan, 1964; Marcallo et al., 1964;

Freire-Maia & Azevedo, 1965; Mi *et al.*, 1965; Fujiki *et al.*, 1966; Krieger, 1966) and resulted in a new theory to discriminate between mechanisms causing the inbreeding effects (Morton *et al.*, 1956). On the whole, inbreeding has been shown to cause an increase in morbidity and mortality, but this effect has been inconstant and variable; for example, Böök found less mortality although more abnormality among the inbred compared to the controls; Schull & Neel (1965) found a difference in the effect of inbreeding upon mortality in Hiroshima compared to Nagasaki; Dolinar (1960) found no recessive traits among the inbred population of Susak; and Fujiki *et al.* found increased mortality in one inbred group but reduced mortality in a neighbouring group; moreover, they found not a single recessive trait among 4694 individuals.

Krieger (1966) has shown that the variability between studies is mainly found among investigators using unselected controls, whereas the studies with sib controls are fairly consistent; nonetheless, there is danger in using a sib control if the propositus is allowed to choose the control, as in the Chicago and Brazilian studies (Slatis *et al.*, 1958; Marcallo *et al.*, 1964) since the sib selected is more likely to be of high social class with healthy children. Even in Schull and Neel's (1965) massive and painstaking collection of data, they were dependent upon the readiness of the Japanese to give reliable answers to their victors, which Haldane, for one, questions (1964).

In this study, the consequences of inbreeding are clear, at least for the 45 couples selected, and there can be no doubt that the detriment, in the main, was due to recessive gene traits because the phenotype and pedigrees were characteristic. But it does not follow that the increased frequency of recessive traits observed throughout the island is due to inbreeding alone. Of the 669 pedigrees that provided sufficient information to permit the identification of close inbreeding, only 51 couples (7·6%) were discovered. This may be an underestimate since Yasuda (1966) has shown that estimates of inbreeding calculated from the observed deviation from Hardy–Weinberg equilibrium for blood groups using maximum likelihood gives approximately twice the inbreeding coefficient calculated from pedigree data. Even if that were so, and the true frequency of cousin marriage was 15% of all marriages, it would still not account for the present findings as can be readily appreciated by considering the consequences of the same level of inbreeding in Europe. If

15% of Europeans married their first cousins the trait frequency of albinism would change in one generation from 1:20,000 to 1:14,000. If the first-cousin rate should be 30%, the trait frequency would only rise to 1:10,000. After several generations of inbreeding at the same level the trait frequency would not measurably alter. Hence it is clear that the observed frequency of one albino for every 852 St. Helenians cannot be attributed to inbreeding alone.

Bernstein (1930) pointed out that the frequency of the recessive zygote aa is given as

$$X^2 + a X (1 - X)$$

where X is the frequency of the recessive gene a and a is the mean coefficient of inbreeding and equivalent to Wright's F. The formula demonstrates that the frequency of the recessive zygote is dependent upon only two parameters, X and a. Since even extreme estimates of a will not account for the observed trait frequencies in these data, the gene frequency must be increased. The causes of this increase will be considered in the next chapter.

In summary, then, inbreeding has only one genetical effect, namely to increase the frequency of the homozygote at the expense of the heterozygote. There are several possible sequelae, some of which were observed:

1. Autosomal recessive traits may increase in frequency. This was seen for deaf mutes, albinos and ichthyosiform erythrodermia.

2. Unknown autosomal recessive traits may appear. Of the several unique findings, those most likely to represent recessive syndromes were the dwarfs with pseudobrachydactyly, the congenital hyperbilirubinaemia and the gross familial unilateral genu valgum.

3. Females homozygous for sex-linked genes may appear. This may have occurred for Christmas disease.

4. Affected males may marry carrier females. This is a possible explanation of the male to male "transmission" in the family with Christmas disease.

5. Mortality may increase due to recessive lethals, causing infertility (no evidence), abortions (no evidence), or stillbirths and infant deaths, which showed a 9·3% increase among the offspring of cousin matings.

6. Mortality may decrease due to the diminished frequency of maternal–foetal incompatibility. There was no evidence in these data, but Einen & Stevenson (1966) have advanced this explanation to account for the reduced frequency of toxaemia of pregnancy among inbred Egyptian women.

7. Exceptionally gifted individuals may increase due to beneficial recessive genes, for which there is no evidence.

8. The variance of quantitative traits may increase as suggested by Wright (1946). This has not been critically examined, but is not noticeable in the analysis of the entire population data. Harwood (1962) reported the same frequency distribution of refractive errors as in England, and Shine et al. (1967) found no increase in variance in blood pressure. These are both crude measures which would not show slight effect, although using more powerful methods no consistent or marked effect of inbreeding on human metrical traits have been observed, and where seen the effect has usually been slight (Morton, 1955; Steinberg, 1963; Mange, 1964; Schull & Neel, 1965; Krieger, 1966).

9. Phenodeviants, that is traits representing the tail of a normal frequency distribution, might increase as proposed by Dempster and Lerner (1950), Edwards (1960), Carter (1961) and Newcombe (1964). There is no evidence for it in these data.

10. Unlinked genes may associate, as suggested by Haldane (1951). Two sibships which manifest different recessive traits were observed, namely ichthyosis and spasticity, and pseudo-brachydactylous dwarfs and spasticity.

11. Morbidity may increase among grandchildren. If there are recessive genes causing non-disjunction in man as in Drosophila (Gowen, 1932), it will be revealed by looking for consanguinity among the grandparents of affected individuals. Penrose (1963) mentioned this possibility, but, I think, has not so far discovered any example.

CHAPTER 6

SYNTHESIS

> We ought not to set them aside with idle thoughts or idle words about curiosities or chances. Not one of them is without meaning; not one that might not become the beginning of excellent knowledge, if only we could answer the question—why is this rare? or being rare, why did it in this instance happen?
>
> (SIR JAMES PAGET, 1882)

THIS study was not originally designed to evaluate the genetical consequences of isolation, indeed it was not designed at all but came about by accident when some rare genetical diseases were encountered. This discovery provoked the question, Are there many rare diseases in St. Helena and, if so, why? The population was then examined to see how many people were normal and how many were abnormal, making detailed records of all abnormalities. It soon became apparent that the local pattern of disease was unusual, and in order to explain the observed departures from expectation, several types of data—genetical, sociological and nutritional, were collected. It was frustrating to find that it was difficult to define expectation due to the problem of sampling errors in small populations and the shortcomings of disease recognition and classification that were listed in Chapter 4. In addition, it turns out that excepting studies on very small communities, there are virtually no other comparable data on congenital and inherited disorders. There are several inbreeding studies that are bigger and better than the present one, but none has adopted such stringent controls and none has discovered new diseases. This is partly because few investigators are fortunate enough to be well-established general practitioners when they are about to begin their research. This study provides proof that, even with a busy and very general practice to run and with limited equipment, an intimate familiarity with the local people and their culture ensures the collection of good-quality data, and it may be that good-quality data cannot be collected otherwise (Haldane, 1963). Far more was achieved in St. Helena than seemed likely when the study started. It was found possible to examine virtually the whole population for a wide range of disease, to obtain extensive pedigree information which could be stringently evaluated, to obtain estimates of the state of nutrition, tobacco consumption, the distribution of fertility, body weight and the arterial blood pressure. Such a collection of data would be very difficult to achieve except in an isolated community. It seems to be true that people are nicer when isolated, they generally know more pedigree information and they have a more co-operative attitude to investigators and, finally, the family are generally accessible. The consequent advantage can readily be appreciated by considering the problem of undertaking a similar study in London or even Honolulu.

Not only do isolates present a favourable social setting for research, but they create special environmental and genetical conditions. Natural environmental variations are subdivided into discrete parcels, within which artificial environmental variations are fewer in number and are more restricted in range than is found in large populations. This provides an opportunity for unique ecologies to develop, such as was found in the Galapagos Islands. The unique conditions on each island provoked the development of unique adaptations. Had the islands been joined, it is unlikely that unique conditions would have developed, and unlikely that any adaptations arising would become established, since with random mating they would always tend to be absorbed into the common gene pool. It is this tendency of isolates to alter gene frequencies that is of prime importance to the geneticist, and it is the demonstration of this phenomenon in these data, albeit on a small scale, that provides the most significant result of this investigation.

A great many rare disorders were found, such as Crigler–Najjar-like syndrome, retinitis pigmentosa, ocular torticollis, lithopaedion, albinism, ichthyosiform erythrodermia and Christmas disease; in addition, there were several that were previously undescribed, such as gross unilateral genu valgum, familial St. Helenian fever and dwarfs with pseudobrachydactyly. Comparing the list of observed anomalies with McKusick's catalogue of inherited disease (McKusick, 1966), it is obvious that the majority of known disorders were not represented at all. But when a disorder was represented, it tended to occur at a higher frequency than would be found in large populations. Inbreeding is part of the explanation but it is not enough, for an inbreeding coefficient as high as $0 \cdot 1$ would not

141

account for the high frequency of recessive traits in St. Helena, as was shown in the previous chapter. It is necessary to postulate a high gene frequency in addition. This could occur if there had been high gene frequencies in the founding population or because of subsequent modification of the original gene frequencies by mutation, migration, drift or selection.

THE FOUNDER PRINCIPLE

The term "founder principle" refers to the changes in gene frequency that may occur whenever a colony is founded by a small number of individuals carrying genes that are rare or unrepresentative of the parent population. Consider a Londoner in the year 1666 who was made homeless in the Great Fire and emigrated to St. Helena, as some did. Whatever rare gene he happened to carry (perhaps he was the only Londoner with Christmas disease), as soon as he became one of St. Helena's 100 settlers, his rarity value changed from $1:500,000$ to $1:100$. The best human example of this phenomenon was established by Dean (1963) who traced the ancestors of all South Africans with porphyria variegata to one of the original fifty founder couples. As the trait is now found in approximately 1% of Boers, it is clear that the disease has persisted at the high frequency that occurred in the founders; although this example is sometimes mistakenly attributed to systematic change.

> ...as witnessed by the amazing spread of the porphyria gene in South Africa. This suggests the possibility that under certain circumstances the incidence of recessive as well as dominant neuro-metabolic disorders could rise in the future as a result of the present trend towards greater degrees of outbreeding. [Price-Evans, 1965.]

OUTBREEDING

The recent racial crossing of Africans, Chinese and Europeans to produce the present race of St. Helenians represents maximal outbreeding which would, in theory, be expected to increase recessive gene frequencies slightly, but over one or two generations its effect would be exceedingly small, since the only genes gained are those that would otherwise have been exposed to selection in the homozygous state under inbreeding. However, if a particular gene was not represented in one of the races, hybrids will have a lower gene frequency, depending on the size of the hybridizing races, rather like the founder principle in reverse. Hybridity will also greatly increase the possibility of new genetical combinations and, in this way, may have evolutionary importance.

RANDOM GENETIC DRIFT

Since the progeny of any generation approximates to a random sample of all possible progeny in that generation, when the total number of progeny is small, sampling errors are likely and gene frequencies will therefore fluctuate. This property of small populations was first appreciated by John Gulick (1887), a missionary to Hawaii, who discovered 200 species of snail belonging to eight genera, over a small area in the valleys of Oahu. It seemed to him more likely that chance rather than natural selection had caused the observed luxuriance.

> Again it may happen that by gradual subsidence a large island will be divided into 2 smaller islands, and thus certain species inhabiting the original island may be indiscriminately isolated. If one, or both, of the sections is very small, the probability of exact similarity in all respects disappears, unless the species is wanting in plasticity and variability. . . . The indiscriminate isolation of a small fragment of a species leads directly to a modification of type in the descendants of the isolated fragment, for the character of a single individual (or even the average character of several individuals) seldom if ever represents the average character of the original stock in every respect.

This process of random drift has been investigated and quantified largely by Sewall Wright (1951) who showed that in small populations (where the selective difference is less than half the reciprocal of the effective population size), loss or fixation of alleles readily occurs. Large populations may show random drift if there should be separation into small subisolates although, except for rare genes, gene frequency changes would balance out, having no overall effect. The census reports in St. Helena suggest that the number of married couples was usually, but not always, greater than 200. This calculation does not take into account racial, social and geographical barriers that created subisolates, within which the founder effect and drift undoubtedly occurred. This is obvious when two of the subisolates, Sandy Bay and Levelwood, are considered. They have a combined population of about 550, most of whom can trace their ancestry through multiple inbred lines to a Mr. X, who was born

143

in the middle of the last century. He was a prolific man leaving twenty-two children who have left over 1000 descendants.

MIGRATION

There has been slow, steady emigration from the island over the past 100 years which will have caused some increase in trait and gene frequencies if emigrants were predominantly of the fitter genotype. However, it is unlikely to have influenced the frequencies of genes with little phenotypic effect in the heterozygote, which is the case for most recessive genes.

The effect of migration on human populations is quite unknown. The simplifying assumption that the quality of emigrants and immigrants is equivalent is untested. In general, one would predict that if the island has received few immigrants, any new genes so introduced would rapidly become diluted in the population. Alternatively, it is possible that unusual genotypes would more often go to, or be left on, a remote island. For example, any deaf mute slaves accidently taken on board, would probably be off-loaded at St. Helena (the first port of call), in which case immigrants would exert a large effect relative to their small numbers. On the other hand, immigrants from small to large countries more often tend to be highly qualified. An indication of this has been given by Professor Titmus (*The Times*, 11 April 1967) who estimates that the 100,000 doctors, scientists and engineers who have emigrated to the United States from smaller countries over the past 18 years cost these countries more to train than the entire foreign aid contribution of the United States to the rest of the world over the same period.

MUTATION

Preliminary analysis of the inbreeding data gave the following inbreeding loads and ratios: for congenital malformations $B = 2 \cdot 55$, $B/A = 98 \cdot 37$; for postnatal mortality $B = 1 \cdot 29$, $B/A = 21 \cdot 7$ (Shine, 1966). This high B/A ratio is hard to explain on a balanced load hypothesis and is to be anticipated on a mutational load hypothesis. However, the existence of a high B/A ratio does not lead to the conclusion that mutations occurring in the past ten generations of St. Helenians have made any appreciable contribution to the present gene pool. Moreover, there is no reason to suppose that

St. Helenians are exposed to mutagens of any sort. As mentioned in Chapter 1, geiger counter measurements over several parts of the island have produced very low counts; synthetic or preserved foods are a small component of the diet and agricultural practice conforms to U.K. standards.

NATURAL SELECTION

The conditions that prevailed on St. Helena were those that might be expected to favour the action of selection, since there has been great variation in fertility (Appendix 6) and there have been conditions of exceptionally high mortality that could have led to differential survival of certain genotypes. The population has been repeatedly decimated by epidemics, for example "distemper" in 1743, measles in 1807 and poliomyelitis around 1800; moreover, the slave trade was attended with appalling mortality.

> 2626 Africans were taken off the coast of Africa of whom 334 died on the passage to St. Helena, and of the 2294 placed under charge of the collector of Customs 865 have died, their principle disease being dysentery, variola and pneumonia. [*Blue Book*, 1842.]

More definite evidence of selection in man might come from segregation analysis, it might come from correlation between genotype and disease, or it might come if genotypes were distributed geographically in a manner predicted from some known selective force. The distribution of genetical disease among isolates is unusual in that rarities, particularly recessive rarities, occur, and commonly with high frequency. In a remote village in Haute Savoie with a population of 290, there were individuals with alkaptonuria, phenylketonuria, congenital cataract, idiopathic tetanus, Friedreich's ataxia and deafness (Dodinval & Klein, 1962). In 3 of the 40 small villages in the Veglia Islands, each one has its own separate brand of recessive disease, namely dwarfism, total albinism and spastic paraplegia (Fraser, 1962), and in an isolated Spanish village near Madrid, with a population of 150, polydactyly was almost universal (Barsky, 1958). Different, and for the most part, recessive traits are found in high frequency among different subgroups of the Amish, for example Ellis van Creveld syndrome (McKusick *et al.*, 1964), dwarfs with cartilage hair hypoplasia (McKusick *et al.*, 1965), interstitial pyelonephritis (Perkoff *et al.*, 1958), limb girdle

145

dystrophy (Jackson & Copey, 1961), a new pyruvate kinase deficiency (Bowman & Procopio, 1963), Christmas disease (Ratnoff, 1958), agammaglobulinaemia, ataxia-telangiectasia and phenylketonuria (Martin *et al.*, 1963). In a Sardinian isolate, one village has a high frequency of congenital myopia and another night-blindness (Siniscalco, 1962). Further examples of rare or unique anomalies among isolates are Kuru in the eastern highlands of New Guinea (Alpers & Gajdusek, 1965), amyotrophic lateral sclerosis in the Mariana and Caroline Islands (Kurland & Mulder, 1954), two varieties of total colour blindness occurring in 1:100 of the inhabitants of the Danish island of Für (Franceschetti *et al.*, 1958), Huntington's chorea, Ramsay Hunt syndrome, amyotrophic lateral sclerosis and familial periodic paralysis among the 900 inhabitants of the island of Réunion (Laude, 1964), the world's highest frequency of G-6-PD deficiency among the isolated Kurdish Jews (Szeinberg, 1963), a high frequency of club foot among Hawaiians (Chung, 1966; Stewart, 1951), osteoectasia in Maui (Stemmermann, 1966), retinitis pigmentosa in Tristan da Cunha (Sorsby, 1963), porphyria in Lapland where it occurs in 1:1000 of the population compared to 1·5:100,000 throughout Sweden (Waldenstrom, 1957), Christmas disease in the Tena valley of Switzerland (Moor-Jankowski *et al.*, 1957), von Willebrands disease, which occurs in 1:10 of the population of some of the small islands of Årland (Eriksson, 1960), and Tay Sachs disease, which occurs 100 times more frequently in Ashkenazi Jews than other Jewish or non-Jewish groups (Myrianthopoulos & Aronson, 1966).

However, there is no apparent pattern to the distribution of these traits between countries that would suggest the action of natural selection, and even if one was apparent, from such a general analysis it would be easy to draw the wrong conclusion. The occurrence of retinitis pigmentosa in both St. Helena and Tristan da Cunha is more likely to be due to common ancestors (the St. Helenian founders of Tristan were related to both affected lines) than it is to the common environment of the South Atlantic Ocean. Similarly, the seeming high frequency of albinos near the equator and the low frequency near the poles has been attributed to heterozygous advantage conferred in the tropics; but one can just as easily postulate that it is due to a physiological effect of temperature if the albino locus in man should have temperature specific alleles as in most mammals. This hypothesis could be tested readily by measuring the skin pigmentation of the Eskimo at the beginning and the end of the winter.

Evidence for selection might come from the distribution of genes within rather than between communities, since from a diagnostic point of view, an internal comparison is more reliable. Generally, in man, dominants outnumber recessives twentyfold (Stevenson, 1958), so that it was surprising to find that the reverse obtains in St. Helena. This was due to an increased frequency of recessives as well as a diminished frequency of dominants. The discrepancy is probably too large to ascribe to chance alone, and it is more likely that the virtual absence of deleterious dominants (a single individual with multiple neurofibromatosis was the only one found) suggests that this class of gene, which is more sensitive to change under selection, has therefore been eliminated. The high frequency of recessive traits in most isolates suggests the possibility that high levels of heterozygosity confer advantage in isolates. The advantage may be due to greater resistance to infectious disease, since the only factor common to all isolates is the virulence of their epidemics. This might explain the low incidence of pulmonary tuberculosis in St. Helena despite the widespread negative Mantoux reaction and poor economic conditions. This postulate is supported by a report (Bridges, 1948) that when the Feugans suffered an epidemic of measles it was the pure-bred who died whereas the hybrids survived.

The preceding argument is speculative because no mathematical models can cope with all the complexities of real human populations, moreover, the previous population size is uncertain, the migration size is uncertain and the gene frequencies in the founder population must be unknown. It is easy enough to establish the blood pressure distribution of the islanders and relatively easy to demonstrate an increased frequency of rare genes, but it is hard to explain how these facts came about. Whilst it is agreed that selection, drift, migration and mutation are the forces that cause evolutionary change, their relative importance is argued; in particular, uncertainty remains about the contribution of stochastic processes. From the information available it does not seem reasonable to attribute the present findings in St. Helena to recent mutation; no evidence could be found for a responsible selective agent, and there was no evidence that immigrants and emigrants were genetically dissimilar, although indirect argument carries some conviction that multiple forces

147

operate. However, there is definite evidence of a tenfold increase in cousin marriage compared to England, and definite evidence that this has increased the frequency of recessive traits. It would appear probable that the increased gene frequencies are largely due to long-term inbreeding, i.e. the founder effect with a secondary contribution from random genetic drift, with perhaps some contribution from differential migration.

A further reason why this discussion is speculative is because an average selective advantage of as little as one in a million will be quite effective in most species (Haldane, 1932) yet obviously impossible to detect in man, whereas the environmental determinants of disease, although generally without the same evolutionary significance, are easier to measure because their effects are more immediate and are several orders of magnitude greater.

It was shown in Chapter 1 why the nature nurture dualism is out of date, yet there is a sense in which it is convenient to retain the distinction. In a given situation one may ask whether a particular disease is environmentally determined, meaning, would it disappear if an environmental factor were removed or added? The investigation into the effects of wearing shoes provides a good illustration of this and shows the power of a properly designed experiment which the St. Helenians were conducting on themselves. Since there were few extraneous differences between the shoe-wearers and the barefoot, it was possible to evaluate the deforming effect of shoes on the feet, even using a crude measure of deformity. Under 2% of the barefoot showed hallux valgus compared to 16% in the men and 48% in the women who had worn shoes for more than 60 years.

The investigation into the prevalence and cause of ischaemic heart disease was prompted by the erroneous impression that the St. Helenians were immune to the disease. Although the disease was measured with a fair degree of precision, the study produced little more than a suggestion of the possible aetiology.

With two exceptions, current environmental factors were not found to be responsible for the observed pattern of congenital or genetical disorders. The first is the lady with congenital hyperbilirubinaemia who was able to remain reasonably healthy despite a genetically determined enzyme deficiency. When she was given 25 mg. propylthiouracil t.d.s. for one year, she developed aplastic anaemia and died. While many thiourea derivatives are known to cause agranulocytosis and occasionally aplastic anaemia (Goodman & Gillman, 1966), until this example, aplastic anaemia following the administration of propylthiouracil had not been reported (Trotter, 1962). This family was able to conjugate salicylamide normally, but unable to conjugate bilirubin; thus, it seems reasonable to suppose that had the conjugating mechanisms been entirely normal, the patient would have been unlikely to develop aplastic anaemia.

The second example is provided by the family with familial St. Helenian fever. This disease showed familial aggregation suggesting a genetical cause, but the localized site of the lesion and generalized systemic disturbance suggests an environmental cause which unfortunately resisted discovery, despite a careful search in Sandy Bay for a likely vector: unfortunate, because it would doubtless have been possible to turn the discovery to practical effect. While a genetical understanding of disease, whether through inbreeding studies or other methods, is exciting to a geneticist, it is regrettable that he cannot apply his knowledge to the problems that the physician is concerned with. For example, if it should be shown that the variation in susceptibility to electrocution (Jex-Blake, 1913) was genetically determined, anyone interested in prevention would doubtless seek the aid of an electrician, not a geneticist. At least it is historically true (with two exceptions) that genetics has contributed nothing to the treatment or prevention of human disease. It was a pathologist who was responsible for the elimination of congenital syphilis (Fleming, 1929), an ophthalmologist who discovered the rubella syndrome (Gregg, 1941), a biochemist who first realized how to treat phenylketonuria (Woolf & Vulliamy, 1951), and a physician who discovered the teratogenicity of thalidomide (Lenz, 1961). [Of course, Dr. Lenz is also a geneticist, but this was irrelevant to his discovery.]

The present study decisively confirms the general belief that the offspring of cousin matings have an increased risk of malformation and early death. If this is universally true, then the avoidance of first-cousin marriage would result in a small reduction of genetical disease; small, because few people marry their cousins. This must provide almost the only purely genetical method yet known of preventing this category of disease. However, within a few years this view may be obsolete because it may soon be possible to apply the vast current knowledge of gene structure and function, even

149

though there is no way of doing so apparent at the moment (Crick, 1965). After all, when Rutherford was asked by a B.B.C. interviewer in 1936 whether there were any conceivable practical applications of nuclear physics, he replied, "absolutely none".

APPENDIX I

Age and Sex Structure of Population Sampled for Hallux Valgus, Ischaemic Heart Disease and Congenital and Inherited Disorders Compared with St. Helena 1956 Census

Age group	St. Helena population (age 5–60 +) M (%)	F (%)	Examined for Hallux Valgus M (%)	F (%)
5–9	18·2	15·8	14·8	16·5
10–19	30·1	26·7	30·6	28·0
20–29	9·4	12·2	14·3	11·7
30–39	9·8	11·3	8·8	11·9
40–49	10·6	11·5	11·2	9·7
50–59	10·7	10·4	10·6	9·3
60 +	11·2	12·1	9·7	12·8
Sex ratio Male:female	0·89		1·14	

Age group	St. Helena population (age 15–60 +) M (%)	F (%)	Examined for Ischaemic Heart Disease M (%)	F (%)
15–19	18·4	14·6	18·1	15·9
20–29	14·9	18·1	21·5	17·6
30–39	15·5	16·8	15·8	18·2
40–49	16·7	17·1	14·2	14·6
50–59	16·9	15·4	15·2	14·1
60 +	17·6	18·0	15·1	19·6
Sex ratio Male:female	0·84		1·02	

Age group	St. Helena population (age 0–60 +) M (%)	F (%)	Examined for Congenital and Inherited Disorders M (%)	F (%)
0–4	15·1	12·7	10·0	10·0
5–14	31·1	28·6	30·8	33·9
15–19	9·9	8·6	10·8	8·9
20–29	8·0	10·6	12·9	9·9
30–39	9·0	9·9	9·5	10·2
40–49	9·0	10·0	8·5	7·9
50–59	9·1	9·1	9·1	7·9
60 +	9·5	10·6	8·5	11·0
Sex ratio Male:female	0·92		1·10	

151

APPENDIX 2

ANALYSIS OF WATER SUPPLIES ON ST. HELENA

	Plantation	Briars	Marble Hall	Longwood	Deadwood	Jamestown		
						Filter bed inlet	Tap general hospital	Ships supply
Alkalinity as $CaCO_3$	6	60	28	16	15	80	80	75
Carbonate	6	60	28	16	15	80	80	75
Non-carbonate	14	165	3	10	11	10	10	20
Hardness total	20	225	31	26	26	90	90	95
Iron	0·04	0·30	0·16	0·06	0·03	0·32	0·24	0·90
Zinc	1·0	0·40	0·50	0·40	0·30	A	A	A
Copper ⎫								
Lead ⎬	A	A	A	A	A	A	A	A
Manganese ⎭								
Fluoride	A_1	0·1	A_1	A_1	0·1	0·2	0·2	0·2
Selenium	A_2	A_2	A_2	A_2	A_2	A_2	A_2	A_2

A—Absent. A_1—Less than 0·1. A_2—Less than 0·05.

APPENDIX 3

Analysis of Nutrition Based on the Replies to the Dietary Questionnaire
(Data from 85 Families—207 Adults, 72 School Children and 41 Infants)

	* Quantity for 85 families	Total no. of calories
Bread	1682·7	1,867,392
Flour	219·8	348,709
Cheese	61	117,887
Sugar	581·5	1,050,050
Butter	41·5	149,712
Margarine	129·5	466,321
Milk (fresh)	49 pints	
(tinned)	51 sweet	51,340
	22 unsweet	
Fish (fresh)	461·5	285,095
(tinned)	32·8 tins	28,923
Corned beef	78	82,275
Bacon	53·6	128,451
Fresh meat (goat and pork)	155	110,037
Rice	517·5	858,222
Potatoes	647·5	257,757
Vegetables	340	30,954
Lard	125	523,702
Eggs (no. per week)	637·5	57,460
Fowl (no. per week)	25·8	40,555
Biscuits	39·5	100,668
Jam	68·2	82,225
Peanut butter (oz per week)	236	40,396
Fruit (tin/fresh)	Quantities of tinned fruit were small, and rather vague figures were given, therefore these were omitted from calculations	
Total per week		
Per 286 adults		6,678,130
Per adult/week		23,346
Per adult/day		3335

* Pounds per week unless otherwise stated.

153

APPENDIX 3 (continued)

COMPARISON OF STATED AVERAGE CONSUMPTION AND CALCULATED CONSUMPTION
DERIVED FROM 1960 IMPORT FIGURES

	Stated average consumption per adult/week	Calculated average consumption per adult/week
	(lb)	(lb)
Sugar	2·03	2·46
Flour { Bread 4·06 / Flour 0·77	4·83	4·61
Cheese	0·21	0·23
Fats		
Butter	0·14	0·10
Margarine	0·45	0·55
Lard	0·44	0·47

The protein, fat, carbohydrate and calorie content of the dietary items was calculated according to McCance and Widdowson (1960) and M.R.C. Special Report Series No. 302 (1962) by Platt.

The average consumption per adult was calculated as recommended by Miss T. Schulz.

APPENDIX 4

COMPARISON OF OCCUPATION OF PEOPLE MARRYING COUSINS AND THOSE
MARRYING NON-COUSINS (DATA FROM 120 PEOPLE)

	Consanguineous	Control
Labourers	24	19
Housewives	31	33
Cooks	1	1
Ministers	1	0
Teachers	1	0
Policemen	1	1
Shop assistants	0	1
Traders	0	1
Taxi drivers	0	2
Mechanics	0	1
Masons	0	1
No occupation	1	0
Summary		
Unskilled	56	53
Skilled	4	7

$\chi^2 = 1\cdot46. \ 0\cdot3 > P > 0\cdot2.$

APPENDIX 4 (continued)

COMPARISON OF MENTAL AND PHYSICAL HEALTH OF PEOPLE MARRYING COUSINS AND THOSE MARRYING NON-COUSINS (DATA FROM 118 PEOPLE—59 COUSINS AND 59 CONTROLS)

	Cousins	Control
Mental illness		
Hysteria	1	3
Anxiety neurosis	1	1
Schizophrenia—acute onset (diagnosis uncertain)	1	0
Low intelligence	0	0
Physical illness		
Congenital or inherited		
Deaf	0	1
Accessory auricle	1	1
Cataract (age of onset uncertain)	0	1
Unilateral choroiditis	0	1
Bat ears	1	0
Diabetes	1	0
Christmas disease	1	0
Congenitally stiff elbow	1	0
Cardiovascular system		
Angina pectoris	1	2
Angina pectoris		
(secondary to hypochromic microcytic anaemia)	0	1
Congestive cardiac failure	0	1
Intermittent claudication	0	1
Varicose leg veins	2	4
Auricular fibrillation	1	0
Rheumatic fever	1	1
Respiratory system		
Emphysema	0	1
Bronchial asthma	0	2
Gastro-intestinal system		
Inguinal hernia	1	1
Umbilical hernia	1	0
Appendectomy	1	2
Leptospirosis	0	1
Infectious hepatitis	1	0
"Jaundice"	0	1
Locomotor system		
Rheumatoid arthritis	1	1
Osteo arthritis	1	2
Femur fracture	2	0
Skin		
Chronic eczema	1	0
Chiropompholyx	1	0

APPENDIX 4 (continued)

	Cousins	Control
Others		
Shingles	1	0
Hysterectomy	2	0
Beri Beri	1	0
Thyrotoxicosis	1	0
Peripheral nerve palsy	0	1
St. Helenian cellulitis	0	1
Obesity	0	1

APPENDIX 5

The Offspring of 45 Consanguineous and 90 Control Matings
45 Consanguineous Marriages

Ref. No. (1)	Husband's age at marriage (2)	Wife's age at marriage (3)	Duration of marriage (4)	Relationship (5)	Total no. of children liveborn and stillborn (6)	No. stillborn (7)	No. died (8)	(9)	Cause of death (10)	Congenital abnormalities No. (11)	Congenital abnormalities Description (12)
50	25	18	45	C.1	5	–	5	–		–	
48	21	9	49	C.1'	7	–	7	1	Unknown cause	1	Psychotic. ♂ aged 27 has been in mental hospital for 5 years
47	24	20	13	C.1'	4	–	4	–		1	♂ aged 10. Bilateral ptosis causing torticollis—see p. 118
46		21	41	C.1	4	–	4	3	Two of unknown cause. One aged 43 died of acute pulmonary tuberculosis	–	
45	17	16	44	C.1	5	–	5	–		1	♀ aged 36. Deaf mute and mentally defective
44	23	17	53	C.1	5	–	5	2	Cause unknown	–	

(1)	(2)	(3)	(4)	(5)	(6)	(7)	(8)	(9)	(10)	(11)	(12)
43	20	28	2	C.l′	1	—	1	—		—	
42	20	19	47	C.l′	2	—	2	—		—	
41	33	38	10	C.l′	1	—	1	—		—	
40	17	15	13	C.l′	6	—	6	—		—	
39	25	22	4	C.l′	0	—	0	—	Unknown cause	—	Ichthyosiform erythrodermia ♂ aged 15 ♂ aged 3
38	59	28	2	C.l′	0	—	0	—		—	
36	30	24	2	C.l′	1	—	1	—		—	Low intelligence, spastic and pale fundi ♀ aged 13 ♀ aged 5
35	34	26	17	C.l	3	—	3	—		2	
34	48	35	12	C.l′	0	—	0	—			
33	20	18	17	C.l	5	—	5	1		2	
32	22	18	18	½S	8	—	8	—			
31	24	19	10+	C.l	4	1	3	—	Unknown cause	—	♂ aged 31. Generalized albino
30	21	18	23	C.l	2	—	2	—		—	
29	20	23	11	C.l	4	—	4	—		—	
27	16	17	29	C.l	7	—	7	2		1	
26	18	18	55	C.l	7	1	6	—		—	♂ aged 13. Low intelligence—headmaster considers him ineducable
25	32	25	13	C.l	1	—	1	—		—	
24	19	16	14	C.l	3	—	3	—		—	

APPENDIX 5 (continued)
45 Consanguineous Marriages

Ref. No. (1)	Husband's age at marriage (2)	Wife's age at marriage (3)	Duration of marriage (4)	Relationship (5)	Total no. of children liveborn and stillborn (6)	No. stillborn (7)	No. died (8)	(9)	Cause of death (10)	Congenital abnormality No. (11)	Congenital abnormality Description (12)
23	21	16	5	C.1	3	—	3	—	Congenital heart disease not further specified	—	♀ aged 37 dwarf with brachydactyly
22	27	25	48	C.1'	8	—	8	2		4	♀ aged 31 dwarf with brachydactyly ♀ monozygous twins aged 40 both with low intelligence, talipes and pale fundi
21	24	21	33	C.1	7	—	7	—	Unknown cause.	—	
20	26	18	37	C.1'	7	—	7	3	♂ died at 5/12 in status epilepticus	—	

(1)	(2)	(3)	(4)	(5)	(6)	(7)	(8)	(9)	(10)	(11)	(12)
19	23	19	9	½C.1	5	–	5	1		–	♂ Arthrogryposis N.B.: considered as congenital abnormality, not as child death
18	29	22	43	C.1′	0	–	0	–	Cause unknown	–	
17	26	24	41	C.1′ × 2	4	2	2	1		–	
16	22	17	7	C.1	3	–	3	–	Cause unknown	–	
15	18	14	19	C.1	9	–	9	1		2	♀ aged 16 mentally defective, ♂ aged 15 mentally defective
14	24	21	7	C.1	5	–	5	1	Cause unknown	–	
13	21	20	21	C.1	5	–	5	2	Deaf mutes ♂ aged 3 deaf mute (+ presumed retinitis pigmentosa) ♀ aged 4 deaf mute (+ presumed retinitis pigmentosa)	3	♂ aged 18 deaf ♀ aged 8 deaf mute with presumed retinitis pigmentosa ♀ aged 10 years mentally defective
12	24	21 + ?	14	C.1	3	–	3	–		–	
11	21	17	12	C.1	3	–	3	–		–	
10	38	23	17	C.1′	0	–	0	–		–	
9	21	20	15	C.1	7	3	4	–	Cause unknown	–	
7	22	18	36	C.1′	7	–	7	–		–	
5	15	20	27	C.1′	12	1	11	1		–	
4	13	24	44	C.1	0	–	0	–		–	
3	–	–	–	C.1′	0	–	0	–		–	

APPENDIX 5 *(continued)*

45 Consanguineous Marriages

Ref. No. (1)	Husband's age at marriage (2)	Wife's age at marriage (3)	Duration of marriage (4)	Relationship of control (5)	Total no. of children liveborn and stillborn (6)	No. stillborn (7)	No. died (8)	(9)	Cause of death (10)	Congenital abnormality No. (11)	Congenital abnormality Description (12)
2	36	30	45	C.1 × 2	9	–	9	2	Cause unknown	–	♀ aged 14 deaf mute and blind, presumed retinitis pigmentosa; ♀ aged 21 low intelligence and vocal abnormality. ♂ aged 18 low intelligence and vocal abnormality
1	23	18	36	C.1	8	–	8	–		4	♂ aged 8 low intelligence and vocal abnormality

APPENDIX 5 (continued)

90 Control Matings

Ref. No. (1)	Husband's age at marriage (2)	Wife's age at marriage (3)	Duration of marriage (4)	Relationship of control to test mate (5)	Total no. of children liveborn and stillborn (6)	No. stillborn (7)	No. liveborn (8)	No. died (9)	Cause of death (10)	Congenital abnormality No. (11)	Congenital abnormality Description (12)	Test / Control } Age of control and propositus (13)
50	17	23	50	Sister	6	—	6	1	Cause unknown	—		69♂ 66♀
50	19	21	26	Brother	8	—	8	2	Cause unknown	—		62♂ 45♀ ♂
48	19	23	20	Niece	5	—	5	1	♂ died at 3/12 deeply jaundiced	—		69♂ 40♀
48	15	10	42	Sister	7	—	7	—		—		60♂ 65♀ ♂
47	17	22	14	Sister	1	—	1	—		—		36♂ 30♀
47	17	32	3	Sister	1	—	1	—		—		38♂ 20♀ 40♂
46	20	16	61	Sister	9	—	9	—		—		40♀ ♂ 40♀ ♀

163

APPENDIX 5 (continued)

90 Control Matings

Ref. No. (1)	Husband's age at marriage (2)	Wife's age at marriage (3)	Duration of marriage (4)	Relationship of control to test mate (5)	Total no. of children liveborn and stillborn (6)	No. stillborn (7)	No. liveborn (8)	No. died (9)	Cause of death (10)	Congenital abnormality No. (11)	Congenital abnormality Description (12)	Test { Age of control / Control ∫ and propositus } (13)
46	20	22	46	Sister	6	–	6	–		–		♀78 / ♀70
45	17	18	39	Sister	9	–	9	–		–		♀54 / ♀61
45	27	22	36	Brother	5	–	5	2	Cause unknown	–		♀58 / ♀62
44	18	21	60	Sister	14	1	13	1	Death unspecified	–		♂76 / ♂66
44	20	22	46	Sister	6	–	6	–		–		♀78 / ♀76
43	18	32	11	Brother	6	–	6	–		–		♀43 / ♀30
43	16	20	7	Sister	2	–	2	–		–		♂23 / ♀22
42	19	28	44	Brother	6	–	6	–		–		♂23 / ♂

(1)	(2)	(3)	(4)	(5)	(6)	(7)	(8)	(9)	(10)	(11)	(12)	(13)
42	20	22	46	Sister	6	—	6	—		—		66♀ 78♀
41	19	23	16	Brother	2	—	2	—		—		43♀ 39♂
41	23	28		Self	2	—	2	—		—		48♀ 48♂
40	18	25	11	Sister	2	—	2	—		—		26♀ 29♀
40	18	23	12	First cousin	3	—	3	—		—		26♂ 29♂
39	22	22	5	Brother	2	—	2	—		—		29♀ 27♀
39	22	20	9	Cousin	2	—	2	—		—		26♀ 31♂
38	18	21	19	Son	3	—	3	—		—		61♀ 36♀
38	22	20	9	Cousin	2	—	2	—		—		30♂ 31♂
36	19	22	11	Brother	6	—	6	—		—		31♀ 31♀
36	18	20	2	First cousin	1	—	1	—		—		28♂ 20♂
35	16	21	13	Brother	3	—	3	—		1	♂ epileptic	42♀ 32♀
35	29	32	23	Brother	1	—	1	—		—		50♂ 55♂

165

APPENDIX 5 (continued)

90 Control Matings

Ref. No. (1)	Husband's age at marriage (2)	Wife's age at marriage (3)	Duration of marriage (4)	Relationship of control to test mate (5)	Total no. of children liveborn and stillborn (6)	No. stillborn (7)	No. liveborn (8)	No. died (9)	Cause of death (10)	Congenital abnormality No. (11)	Congenital abnormality Description (12)	Test / Control — Age of control Control and propositus (13)
34	18	23	40	Sister	3	—	3	1	♂ died unknown cause at 14 years	—		64 ♂ / 58 ♀
34	21	25	18	Sister	3	—	3	—		—		47 ♂ / 38 ♀
33	14	16	18	Brother	8	—	8	—		—		36 ♀ / 34 ♀
33	17	21	8	Sister	3	—	3	—		—		36 ♂ / 25 ♂
32	20	25	23	Sister	4	1	3	—		—		34 ♀ / 43 ♀
32	18	28	12	First cousin	1	—	1	—		—		37 ♀ / 26 ♀
31	18	25	22	Brother	7	—	7	—		—		53 ♂ / 47 ♂

(1)	(2)	(3)	(4)	(5)	(6)	(7)	(8)	(9)	(10)	(11)	(12)	(13)
31	18	22	10	Son	8	–	8	1	Unknown cause	–		58 ♂ 32 ♀
30	19	26	12	Sister	4	–	4	–		–		♂ ♂
30	18	22	37	Brother	4	–	4	–		–		44 31
29	17	27	3	Self	1	–	1	–		–		♂ 53 ♀ 39
29	16	19	19	Sister	6	–	6	–		–		34 34 ♂ ♂
27	15	10	42	Brother	7	–	7	–		–		31 34 ♂ ♀
27	21	23	73	Sister	6	–	6	–		–		55 61 ♀ ♀
26	13	21	29	Son	2	–	2	–		–		55 94 ♀ ♂
26	21	21	28	Daughter	2	–	2	–		–		70 50 ♂ ♀
25	20	28	8	Brother	5	–	5	–		–		70 49 ♂ ♂
25	16	21	20	Brother	4	–	4	–		–		38 36 ♀ ♀
24	18	21	13	Sister	3	–	3	–		–		43 36 ♀ ♂
24	16	18	17	Sister	5	–	5	–		–		33 30 ♂ ♂
												30 32 ♀ ♀

APPENDIX 5 (continued)
90 Control Matings

Ref. No. (1)	Husband's age at marriage (2)	Wife's age at marriage (3)	Duration of marriage (4)	Relationship of control to test mate (5)	Total no. of children liveborn and stillborn (6)	No. stillborn (7)	No. liveborn (8)	No. died (9)	Cause of death (10)	Congenital abnormality No. (11)	Congenital abnormality Description (12)	Age of control { Control } and propositus — Test (13)	Age of control { Control } and propositus — Control (13)
23	16	21	2	Sister	1	–	1	–		–		21 ♀	18 ♀
23	18	22	10	Sister	0	–	0	–		–		26 ♀	28 ♀
22	20	22	58	Sister	6	–	6	–		–		75 ♂	78 ♀
22	21	21	45	Brother	7	–	7	1	Unknown cause	–		73 ♂	66 ♂
21	24	20	44	Sister	8	–	8	2	Unknown cause	–		54 ♀	66 ♀
21	18	23	29	Sister	9	–	9	–		–		57 ♀	46 ♀
20	19	22	36	Sister	11	–	11	2	Unknown cause	–		55 ♂	54 ♀
20	20	22	58	Sister	6	–	6	–		–		63 ♂	78 ♀

(1)	(2)	(3)	(4)	(5)	(6)	(7)	(8)	(9)	(10)	(11)	(12)	(13)
19	18	29	14	Sister	2	–	2	1	Unknown cause	–		28 ♀ 32 ♂
19	22	29	19	Cousin	1	–	1	–		–		32 ♀ 40 ♀
18	33	40	6	Sister	1	–	1	–		–		♀ ♀
18	33	40	6	Niece	1	–	1	–		–		69 ♀ 39 ♀
17	33	40	12	Daughter	1	–	1	–		–		♂ ♂
17	26	53	7	Brother	0	–	0	–		–		65 ♂ 39 ♀
16	16	20	8	Brother	2	–	2	–		–		67 ♀ 65 ♂
16	15	25	30	Sister	1	–	1	–		–		♂ ♂
15	26	27	10	Brother	3	–	3	–		–		23 ♂ 28 ♂
15	17	24	30	Sister	5	–	5	1	Unknown cause	1	♂ aged 6 pulmonary stenosis with reversed shunt	♀ ♀ 27 ♂ 26 ♀
14	15	21	30	Cousin	14	–	14	–		–		36 ♂ 57 ♂ ♂ ♂ 32 ♀ 27 ♀
14	17	24	10		5	–	5	1	Unknown cause	1	♂ aged 6 pulmonary stenosis with reversed shunt	♀ ♀ 28 ♀ 45 ♀ ♀ ♀ 31 ♂ 27 ♂
13	26	27	30	Sister	3	–	3	–		–		♂ ♂ 41 ♀ 57 ♂ ♀ ♀

APPENDIX 5 (continued)

90 Control Matings

(1) Ref. No.	(2) Husband's age at marriage	(3) Wife's age at marriage	(4) Duration of marriage	(5) Relationship of control to test mate	(6) Total no. of children liveborn and stillborn	(7) No. stillborn	(8) No. liveborn	(9) No. died	(10) Cause of death	Congenital abnormality (11) No.	Congenital abnormality (12) Description	(13) Test / Control, Age of control and propositus
13	21	26	20	Sister	6	—	6	—		—		42 ♂ / 41 ♀
12	24	48	27	Self	6	—	6	2	Unknown cause	—		75 ♂ / 75 ♀
12	23	22	—	Sister	5	—	5	—		—		72 ♂ / 86 ♂
11	17	21	11	M.Z. twin	1	—	1	—		—		27 ♀ / 27 ♀
11	16	24	15	Sister	7	—	7	1	♀ aged 12 post-operative death; died one day after repair of harelip	1	♂ deaf mute and parasitic twin. Presumed retinitis pigmentosa	33 ♂ / 29 ♀
10	21	35	33	Brother	11	—	11	—		—		52 ♂ / 68 ♂
10	19	21	13	Sister	4	—	4	—		—		55 ♂ / 31 ♀

(1)	(2)	(3)	(4)	(5)	(6)	(7)	(8)	(9)	(10)	(11)	(12)	(13)
9	15	25	8	Sister	1	–	1	–		–		33 ♀ 23 ♀
9	16	20	7	Brother	2	–	2	–		–		35 ♀ 28 ♂
7	21	20	43	Brother	9	–	9	–		–		56 ♀ 60 ♂
7	25	31	35	Sister	9	–	9	–		2	1 ♂ aged 17 unilateral genu valgum. 1 ♂ aged 18 unilateral choroiditis and blindness since birth	51 ♂ 48 ♂ ♀
5	20	25	28	Sister	7	–	7	–		–		47 ♀ 58
5	19	24	23	Sister	12	–	12	2	1 ♀ aged 21 unknown cause, 1 unknown	2	Peripheral muscular dystrophy	42 ♂ 40 ♂ ♀
4	19		20	Sister	5	–	5	–	Unknown cause	2	♀ aged 35 deaf mute with retinitis pigmentosa; ♂ aged 45 deaf and low intelligence	57 ♀ 65 ♂ ♀
4	29	29	20	Sister	5	–	5	1		–		68 ♀ 48
3	21	18	19	Niece	3	–	3	–		–		53 ♀ 53
3			15	Self	6	–	6	–		–		58 ♂ 58 ♂ ♂

171

APPENDIX 5 (continued)
90 Control Matings

Ref. No. (1)	Husband's age at marriage (2)	Wife's age at marriage (3)	Duration of marriage (4)	Relationship of control to test mate (5)	Total no. of children liveborn and stillborn (6)	No. stillborn (7)	No. liveborn (8)	No. died (9)	Cause of death (10)	Congenital abnormality No. (11)	Congenital abnormality Description (12)	Age of control and propositus — Test / Control (13)
2	21	35	9	Self	2	–	2	–		–		75 ♀
2	22	20	46	Sister	7	–	7	1	Unknown cause	–		68 ♀ / 72 ♀
1	20	25	28	Sister	7	–	7	–		–		♂ / 58 ♀ / 50 ♀
1	29	23	31	Brother	4	1	3	–		–		59 ♀ / 59 ♂ / ♂

Key (Cause of death / Description):

- C.1 — First cousin
- C.1' — First cousin once removed
- C.1 × 2 — Double first cousin
- ½C.1 — Half first cousin
- C.1' × 2 — Double first cousin once removed
- ½S — Half sib

172

APPENDIX 6

Variation in Fecundity

Age group of mother	No. of children born																	Total
	(0)	(1)	(2)	(3)	(4)	(5)	(6)	(7)	(8)	(9)	(10)	(11)	(12)	(13)	(14)	(15)	(17)	
15–19	143	29	6	2														
20–29	33	39	44	36	24	7	11	4	1	8								
30–39	15	22	18	31	34	21	20	16	10	14	5							
40–49	15	11	20	13	15	26	12	12	8	5	3	1	1					
50–59	16	20	15	17	21	16	12	11	9	8	4	9	2	3	1		1	
60–69	26	7	12	12	10	15	6	13	7	3	2	6	3	2	1			
70–79	7	11	12	12	8	4	4	2	1	2		3	3	2		1		
80–89	1	5	3	1	3	3	2	2	2	1				1				
90 +						1	1											
TOTAL	256	144	130	124	115	93	68	60	38	41	14	19	9	8	2	1	1	1123

173

REFERENCES

ABERLE, D. F., BRONFENBRENNER, U., HESS, E. H., MILLER, D. R., SCHNEIDER, D. M. and SPUHLER, J. N. (1963) *Am. Anth.* **65,** 253.

ALDRICH, R. A., STEINBERG, A. G. and CAMPBELL, D. C. (1954) *Pediatrics* **13,** 133.

ALLEN, G. (1965) In *Progress in Medical Genetics,* Vol. IV (ed. by A. G. Steinberg and A. G. Bearn), Grune and Stratton, p. 242.

ALPEN, E. L., MANDEL, H. G., RODWELL, V. W. and SMITH, P. K. (1951) *J. Pharmacol. exp. Ther.* **102,** 150.

ALPERS, M. and GAJDUSEK, D. C. (1965) *Am. J. trop. med. Hyg.* **14** (5) 852.

ALWALL, N., LAURELL, C. B. and NILSBY, I. (1946) *Acta med. scand.* **124,** 114.

ARIAS, I. M. (1962) *J. clin. Invest.* **41,** 2233.

ARNER, G. B. L. (1908) Columbia Univ. Studies in History, Economics and Public Law 31, No. 3.

ATA, M., FISHER, O. D. and HOLMAN, C. A. (1965) *Lancet* **i,** 119.

AYCOCK, W. L. and GORDON, J. E. (1941) *Amer. J. med. Sci.* **201,** 450.

BAKER, I. (1964) Personal communication.

BARNETT, C. H. (1962) *J. Anat. (Lond.)* **96** (4) 489.

BARNICOT, N. A. and HARDY, R. H. (1955) *J. Anat. (Lond.)* **89,** 355.

BAROODY, W. G. and SHUGART, R. T. (1956) *Am. J. Med.* **20,** 314.

BARSKY, A. J. (1958) *Congenital Anomalies of the Hand and their Surgical Treatment,* Charles C. Thomas, Springfield, Illinois.

BATESON, W. (1902) *Mendel's Principles of Heredity,* Cambridge.

BEATSON, A. (1816) *Tracts Relative to the Island of St. Helena,* W. Bulmer & Co., London.

BELL, Sir CHARLES (1836) *The Nervous System of the Human Body,* Edinburgh, p. 434.

BELL, JULIA (1951) *The Treasury of Human Inheritance,* Vol. V, Pt. 1, Cambridge University Press.

BEMISS, J. M. (1858) *Trans. Am. med. Ass.* **11,** 319.

BERNSTEIN, F. (1930) *Z. indukt. Abstamm.- u. Vererb Lehre* **56,** 233.

BEUCHLEY, R. W., DRAKE, R. M. and BRESLOW, L. (1958) *Circulation* **18,** 1085.

BILLING, B. H. (1964) Personal communication.

BLACK, J. A., THACKER, C. K. M., LEWIS, H. E. and THOULD, A. K. (1963) *Brit. med. J.* **ii,** 1018.

BLOUNT, W. P. (1937) *J. Bone Jt Surg.* **19,** 1.

Blue Book (1842) Blue Book reports on the past and present state of Her Majesty's colonial possessions.

BLUMBERG, B. S. (1965) *Int. J. Leprosy* **33** (3) 739.

BOINET, E. (1898) *Rev. Med. (Paris)* **18,** 316.

BÖÖK, J. A. (1957) *Ann. hum. Genet.* **21,** 191.

BOS, R. J. (1894) *Biol. Centralbl.* **14,** 75.

BOSTRÖM, H. and WIDSTRÖM, A. (1965) *Acta med. scand.* **172** (2) 239.

BOURRAT, L., BOULLIAT, G., TRILLET, M. and VEDRINNE, J. (1963) *Lyon méd.* **210** (48) 1117.

BOWDOIN, C. D. (1942) *J. med. Ass. Ga.* **31,** 437.

BOWMAN, H. S. and PROCOPIO, F. (1963) *Ann. int. Med.* **58,** 567.

REFERENCES

BRAIN, Sir RUSSELL (1961) *Speech Disorders*, Butterworths, London.
BRIDGES, E. L. (1948) *Uttermost Part of the Earth*, Hodder & Stoughton, London.
British Medical Journal (1952) Editorial **ii,** 204.
British Medical Journal (1964) Editorial **ii,** 1282.
BRONTE-STEWART, B., KEYS, A. and BROCK, J. F. (1955) *Lancet* **ii,** 1103.
BRONTE-STEWART, B. and PICKERING, G. W. (1959) in *Medical Surveys and Clinical Trails* (ed. by L. J. Witts), Oxford University Press.
BROOKE, T. H. (1824) *A History of the Island of St. Helena*, 2nd edn., Publishers to the East India Co.
BROUSMICHE, E. (1887) *Arch. Méd nav.*
BROWN, R. G., DAVIDSON, L. A. G., McKEOWN, T. and WHITFIELD, A. G. W. (1957) *Lancet* **ii,** 1073.
BURRY, H. S. (1957) British Boot, Shoe and Allied Trade Research Association, Research Report No. 147.
BUTTERWORTH, T. and STREAN, L. P. (1962) *Clinical Genodermatology*, Williams & Wilkins Co., Baltimore, p. 23.

CARLETON, R. A., ABELMANN, N. H. and HANCOCK, E. W. (1958) *New Eng. J. Med.* **259,** 1237.
CARTER, C. O. (1961) *Brit. med. Bull.* **17,** 251.
CASEY, J. P. N. (1903) *S. Afr. med. Rec.* **i,** 107.
CHEDIAK, M. M. (1952) *Rev. Hémat.* **7,** 362.
CHILDS, B., SIDBURY, J. B. and MIGEON, C. J. (1959) *Pediatrics* **23,** 903.
CHUNG, C. S. (1966) Personal communication.
CLEAVE, T. L. and CAMPBELL, G. D. (1966) *Diabetes, Coronary Thrombosis and the Saccharine Disease*, John Wright & Sons, Bristol.
COHEN, A. M. (1963) *Am. Heart J.* **65,** 291.
COTTERMAN, C. W. (1952) *Proceedings of the Second Blood Typing Conference of the Purebred Dairy Cattle Association*, Brattleboro, Vermont.
COWAN, M. A. and ALEXANDER, S. (1961) *Med. Press* **245,** 263.
CRAIGMILE, DORIS A. (1953) *Brit. med. J.* **2,** 749.
CRICK, F. H. C. (1965) *Brit. med. Bull.* **21** (3) 186.
CRIGLER, J. F. and NAJJAR, V. A. (1952) *Pediatrics* **10,** 169.
CROSSKEY, W. R. (1965) *Proc. R. ent. Soc. (Lond.)* **34,** 33.
CROWE, F. W., SCHULL, J. V. and NEEL, J. V. (1956) *A Clinical, Pathological and Genetic Study of Multiple Neurofibromatosis*, Charles C. Thomas, Springfield, Illinois.
CRUZ-COOKE, R., ETCHEVERRY, R. and NAGEL, R. (1964) *Lancet* **i,** 697.

DAHLBERG, G. (1948) *Mathematical Methods for Population Genetics*, Interscience, London.
DAMESHEK, W. and SINGER, K. (1941) *Arch. intern. Med.* **67,** 259.
DANIELS, W. B. and GRENNAN, H. A. (1943) *J. Amer. med. Ass.* **122,** 361.
DANIELSSEN, D. C. and BOECK, W. (1848) *Traité de la opedalskhed*, Ballière, Paris.
DANNEEL, R. (1938) *Naturwissenschaften* **26,** 200.
DARWIN, C. R. (1845) *Journal of Researches into the Natural History and Geology of the Countries Visited During the Voyage of H.M.S. Beagle Round the World, Under the Command of Capt. Fitzroy R.N.*, John Murray, London.
DARWIN, C. R. (1876) *The Effects of Cross-and Self-fertilization of Plants in the Vegetable Kingdom*, London.
DARWIN, G. H. (1875) *Jl. R. statist. Soc.*
DAWBER, T. R., KANNEL, W. B., REROTSKIE, N., STOKES, J., KAGAN, A. and GORDON, T. (1959) *Am. J. publ. Hlth.* **49,** 1349.

176

DEAN, G. (1963) *The Porphyrias*, Pitman, London.

DEMPSTER, E. R. and LERNER, I. M. (1950) *Genetics* **35**, 212.

DODINVAL, P. and KLEIN, D. (1962) *J. Génét. hum.* **11** (1) 1.

DOLINAR, Z. (1960) *Ann. hum. Genet.* **24**, 15.

DOLL, W. R. and HILL, A. B. (1956) *Brit. med. J.* **ii**, 1071.

DRINKWATER, H. (1908) *Proc. Roy. Soc. Edinburgh* **28**, 35.

DRONAMRAJU, K. R. and KHAN, MEERA P. (1964) *Acta genet.* (*Basel*) **13**, 21.

DUNN, L. C. (1965) *Heredity and Evolution in Human Populations*, Atheneum, New York.

EDWARDS, J. H. (1960) *Acta genet.* **10**, 63.

EDWARDS, J. H. (1966) Paper read at the 152nd meeting of Genetical Society of Great Britain.

EINEN, M. and STEVENSON, A. C. (1965) Personal communication.

EMSLIE, M. (1939) *Lancet* **ii**, 1260.

ENGELSMEIER, W. (1937) *Z. indukt. Abstamm.- u. Vererb Lehre.*

ENGLE, E. T. and MORTON, D. J. (1931) *J. Bone Jt Surg.* **13**, 311.

EPSTEIN, F. H. (1964) *Am. Heart J.* **67**, 445.

EPSTEIN, F. H. (1965) *J. chron. Dis.* **18**, 735.

ERIKSSON, A. (1960) *Finska LäkSällsk. Handl.* **104** (suppl.) 136.

FALLS, H. (1966) in *Retinal Diseases* (ed. by S. J. Kimura and W. M. Gaygill), Lea & Febiger, Philadelphia.

FARABEE, WILLIAM, C. (1903) *Inheritance of Digital Malformations in Man*, Papers of the Peabody Museum of American Archaeology and Ethnology, Harvard, 1905, III, No. 3; extract of a D.Phil. degree submitted to Harvard in 1903.

FISHER, Sir RONALD A. (1959a) *Smoking, the Cancer Controversy*, Oliver & Boyd, Edinburgh.

FISHER, Sir RONALD A. (1959b) *Best Articles and Stories*, September, p. 60.

FISHMAN, W. H. and GREEN, S. (1955) *J. biol. Chem.* **215** (2) 527.

FITZPATRICK, THOMAS, B. and QUEVEDO, W. C. (1966) *The Metabolic Basis of Inherited Disease* (ed. Stanbury, J. B., Wyngaarden, J. B. and Frederickson, D. S.), McGraw-Hill, Toronto, 2nd edn., p. 327.

FLEMING, Sir ALEXANDER (1929) *Brit. J. exp. Path.* **10**, 226.

FORSHUFVUD, S., SMITH, H. and WASSEN, A. (1961) *Nature* (*Lond.*) **192**, 103.

FORSIUS, H. and ERIKSSON, A. (1962) *Acta Genet. med.* (*Roma*), **2**, 397.

FRANCESCHETTI, A., JAEGER, W., KLEIN, D., OHRT, V. and RICKLI, H. (1958) Paper read at the 18th International Congress of Ophthalmology, Brussels, September 1958.

FRANCIS, T. (1944) *Op. Domo Biol. hered. hum., Kbh.* **169**, 86.

FRASER, G. R. (1962) *J. Génét. hum.* **13**, 32.

FREIBERG, A. H. and SCHROEDER, J. H. (1903) *Am. J. med. Sci.* **126**, 1033.

FREIRE-MAIA, N. and AZEVEDO, J. B. C. (1965) *Proc. G. Mendel Memorial Symp.* (in press).

FRISCHKNECHT, W., BIANCHI, L. and PILLERI, G. (1960) *Helv. paediat. Acta* **15**, 259.

FROGGATT, P. (1960) *Ann. hum. Genet.* **24**, 213.

FUJIKI, N., ISHIMARU, H., TAKENAKA, S., SUGIMOTU, N., YAMAMOTO, M. and MASUDA, M. (1966) Paper read at 3rd International Congress of Human Genetics.

FUKUHARA, T. and SAITO, S. (1963) *Bull. Tokyo med. dent. Univ.* **10** (2) 333.

GALLOWEY, H. L. (1909) *Lancet* **ii**, 271.

GALTON, Sir FRANCIS (1874) *English Men of Science, their Nature and Nurture*, Macmillan, London.

REFERENCES

GARROD, A. E. (1902) *Lancet* **ii,** 1616.

GEAR, J. (1952) *Ann. int. Med.* **37,** 1.

GEDDA, L. (1951) *Studio dei gemelli*, Edizioni Orrizzonte Medico, Rome.

GEYSER, A. (1958) *Report on Tuberculosis Survey in St. Helena*, Government Printer, St. Helena.

GILBERT, A., LERELBOULLET, P. and HERSCHER, M. (1907) *Bull. Soc. Méd. Paris* **24,** 1203.

GOLDSCHMIDT, ELIZABETH, COHEN, T., BLOCH, N., KELETI L. and WARTSKI, S. (1963) in *The Genetics of Migrant and Isolate Populations* (ed. by Elizabeth Goldschmidt), Williams & Wilkins, New York, p. 183.

GOODMAN, L. S. and GILMAN, A. (1966) *The Pharmacological Basis of Therapeutics*, 3rd. edn., Macmillan, New York.

GORLIN, R. J. and PINDBURY, J. J. (1964) *Syndromes of the Head and Neck*, McGraw-Hill, New York.

GOSSE, P. (1938) *St. Helena 1502–1938*, London, Cassell.

GOWEN, J. W. (1932) *Proc. 6th Int. Cong. Genet.* **2,** 69.

GOWERS, W. R. (1902) *Brit. med. J.* **1,** 1323, 89.

GREGG, N. M. (1941) *Trans. ophthal. Soc. Australia* **3,** 35 (1942).

GULICK, J. T. (1905) *Evolution, Racial and Habitidinal, Controlled by Segration*, Publ. Carnegie Institute, **25,** 1.

GUY, L. P. (1942) *Arch. Ophthal.* **28,** 17.

HAGUE, M. R. and HARVALD, B. (1961) *Acta genet (Basel)* **11,** 372.

HAINES, R. W. and McDOUGALL (1954) *Lancet* **ii,** 1308.

HALDANE, J. B. S. (1932) *The Causes of Evolution*, Harper & Bros., London.

HALDANE, J. B. S. (1936) *Erkenntnis* **6,** 346.

HALDANE, J. B. S. (1951) *Ann. hum. Genet.* **15,** 15.

HALDANE, J. B. S. (1963) *Genetics Today*, Proc. XI International Congress of Genetics, **2,** xci, Pergamon, Oxford.

HALDANE, J. B. S. (1964) *J. Genet.* **59,** 87.

HALLGREN, B. (1959) *Acta psychiat. scand.* V, **34,** suppl., 138.

HANDFORTH, J. R. (1950) *Anat. Rec.* **106,** 119.

HANHART, E. (1925) *Arch. Klaus-Stift. Vererb-Forsch.* **1,** 42, 181.

HANSEN, G. A. (1874) *Norsk Mag. Laegevidensk. Kristiana* **3,** R, 4.

HARDISTY, R. M. (1957) *Brit. med. J.* **1,** 1039.

HARDISTY, R., DORMANDY, K. M. and HUTTON, R. A. (1964) *Brit. J. Haemat.* **10,** 371.

HARDY, R. H. and CLAPHAM, J. C. R. (1951) *J. Bone Jt. Surg.* **33B,** 376.

HARWOOD, K. A. (1962) *A Report on Ocular Refractions on St. Helena*, September 1962.

HARRIS, H., HOPKINSON, D. A., ROBSON, ELIZABETH B. and WHITTAKER, MARY (1963) *Ann. hum. Genet.* **26,** 359.

HAWKINS, T. B., MITCHELL, C. L. and HEDRICK, D. W. (1945) *J. Bone Jt Surg.* **27,** 387.

HEBERDEN, W. (1772) *Med. Trans. roy. Coll. Phys. Lond.* **2,** 59.

HELLER, H. (1962) *Ann. int. Med.* **5612,** 171.

HENSEN, A., MATTERN, M. J. and LOELIGER, E. A. (1965) *Thrombos. Diath. haemorrh. (Stuttg.)* **14,** 341.

HERMANSKY, F. and PUDLAK, P. (1959) *Blood* **14,** 162.

HIGASHI, O. (1954) *Tohoku J. exp. Med.* **59,** 315.

HILLMAN, J. W. and JOHNSON J. T. H. (1952) *J. Bone Jt Surg.* **34,** A 211.

HINKLE, L. E., CARVER, S., BENJAMIN, B. R., CHRISTENSON, N. W. and STONE, B. W. (1964) *Arch. environm. Hlth. (Chicago)* **9,** 14.

HOFFMANN, H. (1936) *Z. Orthop.* **65,** 353.

HOFFMAN, P. (1905) *Am. J. orthop. Surg.* **3,** 105.

HOGBEN, L. (1931) *J. Genet.* **25,** 97.

HOLST, J. C. (1952) *Amer. J. Ophthal.* **35,** 1153.

HUBBLE, D. (1965) *Recent Advances in Paediatrics* (Ed. Gairdner), J. & A. Churchill Ltd., London.

HUMPHREY, N. (1956) *A Review of Agriculture and Forestry in the Island of St. Helena,* Crown Agents, 1957.

IKKALA, E. (1960) *Scand. J. clin. Lab. Invest.* **12,** suppl., 46, Helsinki.

JACKSON, C. E. and COPEY, J. H. (1961) *Pediatrics* **28,** 77.

JAMES, C. S. (1939) *Lancet* **ii,** 1390, 26.

JANISCH, H. R. (1885) *Extracts from the St. Helena Records,* Benjamin Grant, St. Helena.

JEX-BLAKE, A. J. (1913) *Lancet* **i,** 1787.

JOHNSTON, O. (1956) *Clin. Orthop.* **8,** 146.

JOLLIFFE, N. and ARCHER, M. (1959) *J. chron. Dis.* **9,** 636.

KALCEV, B. (1963) *E. Afr. med. J.* **40** (2) 47.

KALMUS, H. (1965) *Diagnosis & Genetics of Defective Colour Vision,* Pergamon, Oxford.

KAPLAN, E. B. (1955) *Bull. Hosp. Jt Dis.* **16,** 88.

KEIZER, D. P. R. (1950) *Paris méd.* **40,** 566.

KERR, C. B. (1965) Variation in X linked traits, thesis, Oxford University.

KEYS, A. (1953) *J. Mt. Sinai Hosp.* **20,** 118.

KEYS, A. (1956) *J. chron. Dis.* **4,** 364.

KEYS, A. (1958) *Ann. int. Med.* **48,** 83.

KIESER-NIELSEN, H. (1953) *Ugeskr. Laeg.* **115,** 1130.

KLEINBERG, S. (1932) *Am. J. Surg.* **95,** 75.

KLOPPERS, P. J. (1961) Personal communication.

KNOWLES, F. W. (1953) *Med. J. Aust.* **1,** 579.

KRAHL, P. (1955) *Pract. oto-rhino-laryng.* (*Basel*) **17,** 249.

KRIEGER, H. (1966) Inbreeding Effects in Northeastern Brazil, thesis, University of Hawaii.

KURLAND, L. T. and MULDER, D. W. (1954) *Neurology Minneap.* **4,** 355, 438.

LAKE, N. C. (1952) *The Foot,* Baillière, Tindall & Cox, London.

LAM SIM-FOOK and HODGSON, A. R. (1958) *J. Bone Jt Surg.* **40A,** 1088.

LANCASTER, Sir JAMES (1940) *The Voyages of Sir James Lancaster* 1591–1603, Series II, Vol. LXXXV, issued by the Hakluyt Soc., London, p. 79.

LANCET (1961) *Annotation* **ii,** 1395.

LANGE, J. (1931) *Crime as Destiny, A Study of Criminal Twins,* London.

LARRIEU, M. J., CAEN, J., LELONG, J. C. and BERNARD, J. (1961) *Nouv. Revue fr. Hémat.* **1,** 662.

LAUDE, R. (1964) *Neuropathies Héréditaires à l'île de La Réunion,* thesis, Lille.

LEJEUNE, J. (1965) Personal communication referring to cases briefly reported.

LEJEUNE, J., LAFOURCADE, J., BERGER, R. and TURPIN, R. (1964) *C. R. Acad. Sci.* (*Paris*) **258** (23) 5767.

LENZ, W. (1961) Discussion at Conference in Dusseldorf, published by Lenz, W. & Knapp (1962) *Drsch. Med. Wschr.* **87,** 1232.

LEVINE P., ROBINSON, E., CELANO, M., BRIGGS, O. and FALKINBURG, L. (1955) *Blood* **10,** 1100.

LINDBERG, H. A., BERKSON, D. M., STAMLER, J. and POINDEXTER, A. (1960) *Arch. int. Med.* **106,** 628.

LITTLE, J. A., SHANOFF, H. M., CSIMA, ADELE, REDMOND, SHIRLEY and YANO, RUTH (1965) *Lancet* **i,** 933.

179

REFERENCES

LLOYD-THOMAS, H. G. (1961) *Brit. Heart J.* **23,** 561.
LORD, J. M. (1956) *J. Path. Bact.* **72,** 627.
LOVERIDGE, A. (1958) *Wirebird (St. Helena)* **2,** 356.
LOVERIDGE, A. (1959a) *Wirebird (St. Helena)* **2,** 413.
LOVERIDGE, A. (1959b) *Wirebird (St. Helena)* **2,** 446.
LOVERIDGE, A. (1961) *Wirebird (St. Helena)* **3,** 663.
LOVERIDGE, A. (1962) *Wirebird (St. Helena)* **3,** 704.
LOVERIDGE A. (1963a) *Wirebird (St. Helena)* **3,** 777.
LOVERIDGE, A. (1963b) *Wirebird (St. Helena)* **3,** 329.
LOVERIDGE, A. (1964) *Wirebird (St. Helena)* **4,** 429.
LOWENSTEIN, F. W. (1961) *Lancet* **i,** 389.
LOWENSTEIN, F. W. (1964) *Am. J. clin. Nutr.* **15,** 175.

MACFARLANE, R. G. (1962) Personal communication.
MACLENNAN, R. (1966) *Lancet* **i,** 1398.
MADDOCKS, I. and LOVELL, R. R. (1962) *Brit. med. J.* **1,** 436.
MANGE, A. P. (1964) *Hum. Biol.* **36,** 104.
MANN, G. V., SHAFFER, R. D., ANDERSON, R. S. and SANDSTEAD, H. H. (1964) *J. Atheroscler. Res.* **4,** 289.
MARCALLO, F. A., FREIRE-MALA, H., AZEVEDO, J. E. C. and SIMOES, I. A. (1964) *Ann. hum. Genet.* **27,** 203.
MARTIN, P. H., DAVIS, L. and ASKEN, D. (1963) *J. Indiana med. Ass.* **56,** 997.
MASTER, A. M. and GEFFER, A. J. (1964) *N.Y. St. J. Med.* **64** (23) 2865.
MASTER, A. M. and ROSENFELD, I. (1964) *J. Am. med. Ass.* **190,** 494.
MAYO, C. H. (1908) *Ann. Surg.* **48,** 300.
MCBRIDE, E. D. (1935) *J. Amer. med. Ass.* **105,** 1164.
MCCANCE, R. A. and WIDDOWSON, E. W. (1960) *The Composition of Foods*, M.R.C. Special Report, Series No. 297, H.M.S.O., London.
MCCORMICK, D. W. and BLOUNT, W. P. (1949) *J. Am. med. Ass.* **141,** 449.
MCELVENNY, R. T. (1944) *Q. Bull. NWest. Univ. med. Sch.* **18,** 286.
MCKEOWN, T. and RECORD, R. G. (1951) *Lancet* **i,** 192.
MCKUSICK, V. A. (1961) *Medical Genetics 1958–1960*, C. V. Mosby Co., St. Louis.
MCKUSICK, V. A. (1964a) *Lancet* **i,** 832.
MCKUSICK, V. A. (1964b) *Circulation* **30,** 326.
MCKUSICK, V. A. (1966) *Mendelian Inheritance in Man*, Johns Hopkins University Press.
MCKUSICK, V. A., ELDRIDGE, R., HOSTETLER, J. A., RUANGWIT, U. and EGELAND, J. A. (1965) *Bull. Johns Hopkins Hosp.* **116** (5) 285.
MCKUSICK, V. A., EGELAND, J. A., ELDRIDGE, R. and KRUSEN, D. E. (1964) *Bull. Johns Hopkins Hosp.* **115,** 306.
MCKUSICK, V. A. and RAPPAPORT, S. I. (1962) *Arch. int. Med.* **110,** 144.
MEAD, N. G., LITHGOW, W. C. and SWEENEY, H. J. (1958) *J. Bone Jt Surg.* **40A,** 1285.
MELLIS, J. C. (1875) *St. Helena*, Reeve, London.
MEYERDING, W. H. and UPSHAW, J. E. (1947) *Am. J. Surg.* **74,** 889.
MI, M. P., AZEVEDO, ELIANE, KREIGER, H. and MORTON, N. E. (1965) *Acta genet. (Basel)* **15,** 177.
MOOR-JANKOWSKI, J. K., TRUOG, G. and HUSER, H. L. (1957) *Acta genet. (Basel)* **7,** 597.
MORLEY, A. J. M. (1957) *Brit. med. J.* **2,** 976.
MORLEY, MARY (1956) *J. Bone Jt Surg.* **38B.**
MORRIS, J. N. and CRAWFORD, M. D. (1958) *Brit. med. J.* **2,** 1458.
MORRIS, J. N., HEADY, J. A., RAFFLE, P. A. B., ROBERTS, C. G. and PARKS, J. W. (1953) *Lancet* **ii,** 1053, 1111.

180

MORTON, D. J. (1935) *The Human Foot*, New York.
MORTON, N. E. (1955) *Ann. hum. Genet.* **20**, 125.
MORTON, N. E. (1959) *Am. J. hum. Genet.* **11**, 1.
MORTON, N. E., CHUNG, C. S. and PETERS, H. A. (1963) in *Muscular Dystrophy in Man & Animals*, (ed. by G. H. Bourne and M. N. Golarz), Karger, Basel.
MORTON, N. E., CHUNG, C. S. and MI, M. P. (1967) *Genetics of Interracial Crosses in Hawaii*, Karger, Basel.
MORTON, N. E., CROWE, J. F. and MULLER, H. J. (1956) *Proc. nat. Acad. Sci. (Wash.)* **42**, 855.
MUNDY, P. (1698) *The Travels of Peter Mundy*, London.
MYRIANTHOPOULOS, N. C. and ARONSON, S. M. (1966) *Am. J. hum. Genet.* **18**, 313.

NEEL, J. V. (1958) *Am. J. hum. Genet.* **10**, 398.
NEW YORK HEART ASSOCIATION CRITERIA COMMITTEE (1964) *Diseases of the Heart and Blood Vessels*, 6th edn., J. & A. Churchill Ltd., London.
NEWCOMBE, H. B. (1964) in *Proc. 2nd Int. Conf. on Congenital Malformations N.Y.* 1963, International Medical Congress, New York.
NEWMAN, H. H. and FREEMAN, F. N. (1937) *Twins, A Study of Heredity and Environment*, University of Chicago Press, Chicago.
NISSEN, K. I. (1947) *Proc. R. Soc. Med.* **40**, 923.
NORRIS, T. (1958) *Review of Food and Nutrition on the Island of St. Helena*, Government Printing Office, St. Helena.

OLIVER, M. F. and STUART-HARRIS, C. H. (1965) *Brit. med. J.* **2**, 1203.
OSLER, W. (1910) *Lancet* **i**, 839.
OSTER, J. (1953) *Op. Domo Biol. hered. hum., Kbh.* **32**, 1.
OWEN, C. A., AMUNDSEN, M. A., THOMPSON, J. H., JR., STITTELL, J. A., BOWIE, E. J. W., STILWELL, G. G., HEWLETT, J. S., MILLS, S. D., SAVER, W. G. and GAGE, R. P. (1964) *Am. J. Med.* **37**, 71.

PAPP, O. A., PADILLA, L. and JOHNSON, A. L. (1965) *Lancet* **ii**, 259.
PASMA, A. and WILDERVANCK, L. S. (1956) *Overgedrukt uit archivum chrurgicum neerlandicum*, Vol. VIII, Fasc. 1, p. 43.
PENROSE, L. S. (1934) *The Influence of Heredity on Disease*, Lewis, London.
PENROSE, L. S. (1956) *Folia hered. path (Milano)*, Estratto V, Fasc. 1, p. 79.
PENROSE, L. S. (1963) *The Biology of Mental Defect*, Sedwick & Jackson Ltd., London, 3rd ed.
PERKOFF, G. T., STEVENS, F. E., DOLOWITZ, D. A. and TYLER, F. H. (1958) *Arch. int. med.* **102**, 733.
PITTMAN, M. A. and GAHAM, J. B. (1964) *Am. J. med. Sci.* **247**, 293.
PLATT, B. S. (1962) M.R.C. Special Report Series, *Tables of Representative Values of Foods Commonly Used in Tropical Countries*, H.M.S.O., London.
PLOTT, D. (1964) *New Eng. J. Med.* **271** (12) 593.
POPPER, Sir KARL R. (1963) *Conjectures and Refutations*, Basic Books, New York, p. 49.
PRICE-EVANS, D. A. (1965) in *Biochemical Approaches to Mental Handicap in Children* (ed. by J. D. Allan and K. S. Holt), Williams & Wilkins, New York.
PYKE, D. A. (1963) *Proc. R. Soc. Med.* **56**, 567.

QUICK, A. J. (1957) *Hemorrhagic Diseases*, Lea & Febiger, Philadelphia.
QUICK, A. J. and HUSSEY, C. V. (1963) *Am. J. med. Sci.* **245** (5) 643.

RAMAZZINI, B. (1700) *De morbis artificium diatriba mutinae* (translated by R. James, in *A Dissertation on Endemial Diseases* by F. Hoffman, London, 1746).

REFERENCES

Ramgren, O. (1962) *Acta med. scand.* **379.**

Ratnoff, O. D. (1958) *Advanc. int. Med.* **9,** 107.

Record, R. G. and Whitfield, A. G. (1964) *Brit. J. prev. soc. Med.* **18,** 202.

Rimoin, D., Merimee, T. J. and McKusick, V. A. (1966) Paper read at 3rd International Congress of Human Genetics.

Roberts, D. F. (1957) Med. Res. Coun. R.N.P. 57/895, H.M.S.O., London.

Rogers, W. R. and Hurst, W. D. (1964) *Northw. Med. (Seattle)* **63,** 702.

Rosenman, R. H. and Friedman, M. (1963) *J. Am. med. Ass.* **184,** 934.

Russek, H. I. (1964) *Geriatrics* **19** (6) 425.

Sandelin, T. (1922) *Finska LäkSällsk. Handl.* **64,** 543.

Schmid, R. (1966) in *The Metabolic Basis of Inherited Disease* (ed. Stanbury, J. B.), McGraw-Hill, Toronto, p. 890.

Schoental, R. (1960) *Proc. R. Soc. Med.* **53,** 284.

Schrire, V. (1958) *Am. Heart J.* **56,** 280.

Schrire, V. (1964) *S. Afr. med. J.* **38,** 46.

Schull, W. J. and Neel, J. V. (1958) *Am. J. hum. Genet.* **10,** 294.

Schull, W. J. and Neel, J. V. (1965) *The Effects of Inbreeding on Japanese Children,* Harper & Row, New York.

Schwartz, D., Lellouch, G., Angurera, G., Beaumont, J. L. and Lenegre, J. (1966) *J. chron. Dis.* **19,** 35.

Shine, I. B. (1964) *Brit. J. Derm.* **76,** 357.

Shine, I. B. (1965) *Brit. med. J.* **1,** 1648.

Shine, I. B. (1966a) Paper read at 3rd International Congress of Human Genetics, Chicago.

Shine, I. B. (1966b) *Lancet* **ii,** 278.

Shine, I. B. and Allison, P. R. (1966) *Lancet* **i,** 951.

Shine, I. B. and Barr, A. (1966) unpublished results.

Shine, I. B. and Corney, G. (1966) *J. med. Genet.* **3,** 124.

Shine, I. B., Morton, N. E. and Parke, M. (1967) *Am. J. hum. Genet.* (in press).

Shine, I. B. and Morton, N. E. (1966) (in preparation).

Shine, I. B., Morton, N. E. and Parke, M. (1967) (in preparation).

Simpson, N. E. G. and Biggs, R. (1962) *Brit. J. Haemat.* **8,** 191.

Sinclair, H. M. (1959) *The Assessment of Nutriture in Medical Surveys and Clinical Trials* (ed. L. J. Witts), Oxford University Press.

Siniscalco, M. (1962) *The Genetics of Migrant and Isolate Populations* (ed. Elizabeth Goldschmidt), Williams & Wilkins, p. 203.

Sjørgren, T. and Larsson, T. (1949) *Acta psychiat. (Kbh.),* Suppl. 56, 49.

Sjørgren, T. and Larsson, T. (1957) *Acta psychiat. (Kbh.),* Suppl. 113.

Slater, E. (1953) *Psychotic and Neurotic Illness in Twins,* Medical Research Council (Lond.) Special Rep., Series 278, H.M.S.O., London.

Slatis, H. M., Reis, R. H. and Hoene, R. E. (1958) *Am. J. hum. Genet.* **10,** 446.

Smithells, R. W. (1966) *Lancet* **i,** 1.

Snyder, L. (1959) *Science* **129,** 1.

Sohar, E., Heller, J., Pras, M. and Heller, H. (1961) *Arch. int. Med.* **107,** 529.

Somers, K. and Rankin, A. M. (1962) *Brit. Heart J.* **24,** 542.

Sorsby, A. (1958) *Brit. med. J.* **2,** 1587.

Sorsby, A. (1963) *Trans. R. Soc. trop. Med. Hyg.* **57,** 15.

Spickett, S. G. (1962) *Leprosy Rev.* **33** (2) 76.

Spuhler, J. N. (1965) quoted in Schull and Neel's *The Effects of Inbreeding on Japanese Children* (ed. by Schull, W. J. and Neel, J. V.), Harper & Row, New York, p. 315.

Steinberg, A. G. (1963) in *The Genetics of Migrant and Isolate Populations* (ed. by Elizabeth Goldschmidt), Williams & Wilkins, New York.

182

STEMMERMANN, G. N. (1966) *Am. J. Path.* **48**, 661.
STERN, C. (1960) *Principles of Human Genetics*, W. H. Freeman & Co., San Francisco.
STEVENSON, A. C. (1953) *Ann. hum. Genet.* **18**, 50.
STEVENSON, A. C. (1958) United Nations Document A/AC 82/9/R.104, New York, partially reproduced in *Report of the Scientific Committee on the Effects of Atomic Radiation*, Suppl. No. 17, A/3838, p. 197.
STEVENSON, A. C. (1963) in *Exposure of Man to Radiation in Nuclear Warfare* (ed. by Rust, J. H. and Méwissen, D. J.), Elsevier Publ. Co., Amsterdam.
STEVENSON, A. C. and CHEESEMAN, E. A. (1956) *Ann. hum. Genet.* **20**, 177.
STEVENSON, A. C., JOHNSTON, H. A., GOLDING, D. R. and STEWART, M. I. P. (1966) *Comparative Study of Congenital Malformations* and Suppl. to Vol. 34 of W.H.O. Bulletin.
STEWART, S. F. (1951) *J. Bone Jt Surg.* **33A**, 557.
STOLLER, A. and COLLMANN, R. D. (1965) *Nature* **208**, 903.
STONEHOUSE, B. (1960) *Wideawake Island: The Story of the B.O.U. Centenary Expedition to Ascension*, London, Hutchinson.
SUTTER, J. and TABAH, L. (1951) *Population* **6**, 59.
SUTTER, J. and TABAH, L. (1952) *Population* **7**, 249.
SUTTER, J. and TABAH, L. (1957) *Cold Spr. Harb. Symp. quant. Biol.* **22**, 374.
SUTTON, R. L. (1956) *Diseases of the Skin*, C. V. Mosby & Co., St. Louis.
SZABÖ, L. and EBREY, P. (1963) *Acta paediat. Acad. sci. hung.* **4**, 153.
SZEINBERG, A. (1963) in *The Genetics of Migrant and Isolate Populations* (ed. by Elizabeth Goldschmidt), Williams & Wilkins, New York, p. 69.

TANAKA, K. (1963) in *The Genetics of Migrant and Isolate Populations* (ed. by Elizabeth Goldschmidt), Williams & Wilkins, New York, p. 169.
TATE, N. (1963) *Lancet* **ii**, 1090.
TATLOCK, H. (1947) *J. clin. Invest.* **26**, 287.
TEMTAMY, S. A. (1965) Genetic Factors on Hand Malformations. Ph.D. Thesis, Johns Hopkins Hospital.
TROTTER, W. R. (1962) *Diseases of the Thyroid*, Blackwell Sci. Publ., Oxford.
TRUSLOW, W. (1925) *J. Bone Jt Surg.* **7**, 98.
TRYGSTAD, O. and SEIP, M. (1964) *Acta paediat.* (*Uppsala*) **53**, 527.
TURNER, R. (1962) *Bull. trimest. d'information des domains français de Sainte-Hélène* **7**.
TURPIN, R. and LEJEUNE, J. (1965) *Les Chromosomes humains* (Carotype normal et variation pathologique), Masson, Paris.
TYLOR, Sir EDWARD B. (1865) *Researches into the Early History of Mankind and Development of Civilisation*, London, reprinted 1964, University of Chicago Press, Chicago.

VAN DER BOSCH, J. (1959) *Ann. hum. Genet.* **23**, 91.
VENNING, P. (1956) *Am. J. Phys. Anthrop.* **14** (2) 129.
VERLOOP, M. C., VAN WIERINGEN, A., VUYLSTEKE, J., HART, H. C. and HUIZINGA, J. (1964) *Med. Klin.* **59**, 408.

WALDENSTROM, J. (1957) *Amer. J. Med.* **22**, 758.
WALLACE, A. R. (1880) *Island Life*, Macmillan, London.
WALTON, J. N. (1963) in *Muscular Dystrophy in Man and Animals*, Karger, Basel (ed. Bourne, C. H. and Gollancz, M. N.).
WALTON, J. N. and NATTRASS, F. J. (1954) *Brain* **77**, 169.
WEISSMAN, S. L., KHERMOSH, O. and ADAM, A. (1963) in *The Genetics of Migrant and Isolate Populations* (ed. by Elizabeth Goldschmidt), Williams & Wilkins, New York, p. 313.
WELANDER, L. (1951) *Acta med. scand.*, Suppl. 265, **141**, 1.

REFERENCES

WELLS, L. H. (1931) *Am. J. phys. Anthrop.* **15,** 185.
WELLS, R. S. and KERR, C. B. (1966) *Brit. med. J.* **1,** 947.
WILKINSON, P. B. (1936) *Report on Beri Beri,* Govt. Printing Office, St. Helena.
WILKINSON, P. B. (1938) *The Caduceus,* University of Hong Kong, **17,** 4, 11, 206.
WITTERT, Baron VON F. (1607) *Journal de l'amiral Wittert 1607–1610,* Liège [Brussels printed], 1875.
WOOD, P. (1958) *Diseases of the Heart and Circulation,* 2nd edn., Eyre & Spottiswoode, London.
WOOLFE, C. M. and GRANT, R. B. (1962) *Am. J. hum. Genet.* **14,** 391.
WOOLF, L. and VULLIAMY, D. G. (1951) *Arch. Dis. Childh.* **26,** 487.
WRIGHT, S. (1922) *Am. Nat.* **56,** 330.
WRIGHT, S. (1951) *Ann. Eugen. (Lond.)* **15,** 323.
WRIGHT, S. (1950) *J. cell. comp. Physiol.,* Suppl. **35,** 187.
WYNNE-DAVIES, RUTH (1965) *J. med. Genet.* **2,** 227.

YASUDA, N. (1966) The Genetical Structure of Northeastern Brazil, Ph.D. Thesis, University of Hawaii.
YUDKIN, J. (1957) *Lancet* **ii,** 155.
YUDKIN, J. and RODDY, S. (1964) *Lancet* **ii,** 6.
YUDKIN, J. (1964) *Lancet* **ii,** 4.
YUDKIN, J. (1965) *Lancet* **i,** 1218.

ZYCHOWICZ, C. (1964) *Pol. Tyg. lek.* **19/20,** 745.

INDEX